Environmental Analysis

ANALYTICAL CHEMISTRY BY OPEN LEARNING

ACOL (Analytical Chemistry by Open Learning) is a well established series which comprises 32 open learning books and 8 computer based training packages. This open learning material covers all the important techniques and fundamental principles of analytical chemistry.

Books

Samples and Standards
Sample Pretreatment
Classical Methods Vols I and II
Measurement, Statistics and Computation
Using Literature
Instrumentation
Chromatographic Separations
Gas Chromatography
High Performance Liquid Chromatography
Electrophoresis
Thin Layer Chromatography
Visible and Ultraviolet Spectroscopy
Fluorescence and Phosphorescence
Infrared Spectroscopy
Atomic Absorption and Emission Spectroscopy
Nuclear Magnetic Resonance Spectroscopy

X-ray Methods
Mass Spectrometry
Scanning Electron Microscopy and Microanalysis
Principles of Electroanalytical Methods
Potentiometry and Ion Selective Electrodes
Polarography and Voltammetric Methods
Radiochemical Methods
Clinical specimens
Diagnostic Enzymology
Quantitative Bioassay
Assessment and Control of Biochemical Methods
Thermal Methods
Microprocessor Applications
Chemometrics
Environmental Analysis

Software

Atomic Absorption Spectroscopy
High Performance Liquid Chromatography
Polarography
Radiochemistry
Gas Chromatography
Fluorescence
Quantitative IR-UV
Chromatography

Further information: ACOL Office
University of Greenwich,
Avery Hill Road,
Eltham,
London
SE9 2HP

Environmental Analysis

Analytical Chemistry by Open Learning

Author:
ROGER N. REEVE
University of Sunderland

Editor:
JOHN D. BARNES
University of Greenwich

Published on behalf of ACOL (University of Greenwich)
by
JOHN WILEY & SONS
Chichester · New York · Brisbane · Toronto · Singapore

Copyright © 1994 University of Sunderland

Published by John Wiley & Sons Ltd,
        Baffins Lane, Chichester,
        West Sussex PO19 1UD, England
                Telephone National   Chichester (0243)779777
                        International   +44 243 779777

Other Wiley Editorial Offices

John Wiley & Sons, Inc., 605 Third Avenue,
New York, NY 10158-0012, USA

Jacaranda Wiley Ltd, 33 Park Road, Milton,
Queensland 4064, Australia

John Wiley & Sons (Canada) Ltd, 22 Worcester Road,
Rexdale, Ontario M9W 1L1, Canada

John Wiley & Sons (SEA) Pte Ltd, 37 Jalan Pemimpin #05-04,
Block B, Union Industrial Building, Singapore 2057

British Library Cataloguing in Publication Data

A catalogue record for this book is available from the British Library

ISBN 0 471 95134 X (cloth)

ISBN 0 471 93833 5 (paper)

Typeset in 11/13 Times by Techset Composition Ltd, Salisbury
Printed and bound in Great Britain by Biddles, Guildford, Surrey

 **THE ACOL PROJECT**

This series of easy to read books has been written by some of the foremost lecturers in Analytical Chemistry in the United Kingdom. These books are designed for training, continuing education and updating of all technical staff concerned with Analytical Chemistry.

These books are for those interested in Analytical Chemistry and instrumental techniques who wish to study in a more flexible way than traditional institute attendance or to augment such attendance.

ACOL also supply a range of training packages which contain computer software together with the relevant ACOL book(s). The software teaches competence in the laboratory by providing experience of decision making in the laboratory based on the simulation of instrumental output while the books cover the requisite underpinning knowledge.

The Royal Society of Chemistry uses ACOL material to run a regular series of courses based on distance learning and residential workshops.

Further information on all ACOL materials and courses may be obtained from:

ACOL-BIOTOL Office, University of Greenwich, Avery Hill Road, London SE9 2HB. Tel: 081-316 9600 Fax: 081-316 9672.

How to Use an Open Learning Book

Open Learning books are designed as a convenient and flexible way of studying for people who, for a variety of reasons, cannot use conventional education courses. You will learn from this book the principles of one subject in Analytical Chemistry, but only by putting this knowledge into practice, under professional supervision, will you gain a full understanding of the analytical techniques described.

To achieve the full benefit from an open learning book you need to plan your place and time of study.

- Find the most suitable place to study where you can work without disturbance.

- If you have a tutor supervising your study discuss with him, or her, the date by which you should have completed this text.

- Some people study perfectly well in irregular bursts; however, most students find that setting aside a certain number of hours each day is the most satisfactory method. It is for you to decide which pattern of study suits you best.

- If you decide to study for several hours at once, take short breaks of five or ten minutes every half hour or so. You will find that this method maintains a higher overall level of concentration.

Before you begin a detailed reading of the book, familiarize yourself with the general layout of the material. Have a look at the course contents list at the front of the book and flip through the pages to get a general impression of the way the subject is dealt with. You will find that there is space on the pages to make comments alongside the text as you study—your own notes for highlighting points that you feel are particularly important. Indicate in the margin the points you would like to discuss further with a tutor or fellow student. When you come to revise, these personal study notes will be very useful.

Π   When you find a paragraph in the book marked with a symbol such as is shown here, this is where you get involved. At this point you are directed to do things: draw graphs, answer questions, perform calculations, etc. Do make an attempt at these activities. If necessary cover the succeeding response with a piece of paper until you are ready to read on. This is an opportunity for you to learn by participating in the subject and although the text continues by discussing your response, there is no better way to learn than by working things out for yourself.

We have introduced self-assessment questions (SAQs) at appropriate places in the book. These SAQs provide for you a way of finding out if you understand what you have just been studying. There is space on the page for your answer and for any comments you want to add after reading the author's response. You will find the author's response to each SAQ at the end of each part of the book. Compare what you have written with the response provided and read the discussion and advice.

At intervals in the text you will find a Summary and List of Objectives. The Summary will emphasise the important points covered by the material you have just read and the Objectives will give you a checklist of tasks you should then be able to achieve.

You can revise the Unit, perhaps for a formal examination, by re-reading the Summary and the Objectives, and by working through some of the SAQs. This should quickly alert you to areas of the text that need further study.

At the end of the book you will find an index and also a Periodic Table.

Contents

Study Guide

This Unit has been written to introduce you to the application of analytical chemistry to environmental problems. It is written with an assumption of a basic knowledge of common analytical techniques as would be acquired, for instance, during the first two years of an undergraduate degree programme in chemistry. More specialised techniques which would not be found in more general analytical books are described within the text. Little more than a superficial knowledge of the environment is assumed although an interest to learn about the subject is essential.

For those wishing to revise their knowledge of basic analytical techniques prior to reading sections of this Unit, a list of suitable learning material is given at the end of the Study Guide.

Part 1 of *Environmental Analysis* introduces you to simple concepts necessary in the study of the environment, to what we mean by the term 'pollution' and to the role of analytical chemistry. This theme develops in Part 2 to a discussion of pollutant dispersion, reconcentration and final degradation, important concepts to understand when setting up an analytical monitoring scheme.

The remaining six parts of the Unit cover the analysis of water, solid, and atmospheric samples. The techniques discussed develop in complexity, starting with simple volumetric techniques for water quality measurements, and finishing with a discussion of ultra-trace analysis, where determinations are made close to the current limits of detection. Parts 3 and 4 are devoted to the analysis of water samples. Part 5 is concerned with the analysis of solids (biological materials, soils and sediments), particularly with respect to sample preparation. Parts 6 and 7 extend the discussion to the analysis of gaseous and particulate components of the atmosphere. Part 8 is concerned with the special problems of ultra-trace analysis.

PREKNOWLEDGE

Suitable depth of treatment of analytical methods may be found in many standard analytical textbooks, e.g.

(a) J. H. Kennedy, *Analytical Chemistry: Principles*, 2nd edn, Saunders College Publishing, 1990.

(b) D. A. Skoog, D. M. West and F. J. Holler, *Analytical Chemistry: An Introduction*, 5th edn, Saunders College Publishing, 1990.

For those wishing to study the background subject areas in more detail, using open learning material, other texts within the ACOL series would be appropriate. In particular the following would be found useful, listed with the parts of this Unit where the material is introduced.

Parts 1 and 2 onwards:

Samples and Standards

Sample Pretreatment and Separation

Part 3 onwards:

Classical Methods, Volume 1

Visible and Ultraviolet Spectroscopy

Atomic Absorption and Emission Spectroscopy

High Performance Liquid Chromatography

Part 4 onwards:

Gas Chromatography

Chromatographic Separations

Supporting Practical Work

1. GENERAL CONSIDERATIONS

The practicals illustrate some of the wide range of methods which may be used for environmental analysis. They require simple equipment available in most analytical laboratories. The first group of experiments need volumetric glassware, a visible spectrometer and a flame photometer. Two other experiments are then suggested for consideration if a gas chromatography and an atomic absorption spectrometer are available.

2. AIMS

(a) To provide practical experience in the analysis of liquid, solid and gaseous environmental samples.

(b) To illustrate the application of both classical (volumetric) and instrumental techniques to environmental analysis.

(c) To demonstrate simple extraction procedures for liquid, solid and gaseous samples.

3. SUGGESTED EXPERIMENTS

(a) Determination of the total hardness of water by EDTA titration.

(b) Determination of the dissolved oxygen concentration in water by a Winkler titration.

(c) The determination of phosphate in water by visible spectrometry after formation of a phosphomolybdate complex.

(d) The determination of calcium and potassium in leaves and soils by emission spectrometry (flame photometry).

(e) The determination of nitrogen dioxide in the atmosphere using diffusion tubes and spectrometric analysis.

If a gas chromatograph and an atomic absorption spectrometer are available then the following experiments could be carried out:

(f) The determination of halocarbons in water after solvent extraction.

(g) The determination of magnesium in tap water.

Bibliography

1. TEXTBOOKS ON ENVIRONMENTAL ANALYSIS

(*a*) I. L. Marr and M. S. Cresser, *Environmental Chemical Analysis*, International Textbook Company, 1983.

(*b*) S. E. Allen (Ed.), *Chemical Analysis of Ecological Materials*, 2nd edn, Blackwell Scientific Publications, 1989.

(*c*) W. Fresenius, K. E. Quentin and W. Scheider (Eds), *Water Analysis*, Springer-Verlag, 1988.

(*d*) J. P. Lodge, Jr (Ed.), *Methods of Air Sampling and Analysis*, 3rd edn, Lewis Publishers, 1989.

Reference (a) is a student text covering all areas of environmental analysis. References (b)–(d) have introductory sections describing commonly used instrumentation and techniques followed by detailed methods for specific analytes.

2. COMPILATIONS OF STANDARD METHODS

These have been produced by environmental authorities worldwide and include:

(*a*) *Methods for the Examination of Waters and Associated Materials*, Standing Committee of Analysts, HMSO, London.

(*b*) *Methods for the Determination of Hazardous Substances*, Health and Safety Executive, HMSO, London.

(c) US Environmental Protection Agency (EPA) Methods, including

500 Series—Drinking Water

600 Series—Waste Water

NTIS, Springfield, Virginia

(d) *Manual of Analytical Methods*, National Institute of Occupational Safety and Health (NIOSH), NTIS, Springfield, Virginia

3.   TEXTBOOKS ON ENVIRONMENTAL CHEMISTRY

(a) P. O'Neil, *Environmental Chemistry*, 2nd Edn, Chapman and Hall, 1993.

(b) S. E. Manahan, *Environmental Chemistry*, 5th Edn, Lewis Publishers, 1991.

Acknowledgements

Figure 1.3b is taken from *Official Journal of the European Communities*, 30.8.80.

Figure 2.2a is taken from P. O'Neil, *Environmental Chemistry*, 2nd Edn, Chapman and Hall, 1993.

Figures 2.3b and 2.3c are taken from C. T. Chiou, V. H. Freed, D. W. Schnedding and R. L. Kohnert, *Environmental Science and Technology*, **11**, 475–478 (1978).

Figures 2.3g, 3.1b and 3.2a have been drawn by Ms A. T. Smith

Figure 3.1a is taken from R. J. Gibbs, *Science*, **170**, 1088–1090 (1970).

Figure 4.2d is taken from *Gas Chromatographic and Associated Methods for the Characterisation of Oils, Fats, Waxes and Tars 1982, Methods for the Examination of Waters and Associated Materials*, HMSO.

Figure 4.3e is taken from *Dionex Technical Note TN25*, Dionex (UK) Ltd

Figure 5.5a is taken from the *The Sampling and Initial Preparation of Sewage and Waterworks Sludges, Soils, Sediments and Plant Materials Prior to Analysis 1977*, Methods for the Examination of Waters and Associated Materials, HMSO.

Figure 6.2b is taken from *British Standard* BS 1747 Pt 3: 1969, BSI Standards.

Figure 7.2a is taken from *General Methods for the Gravimetric Determination of Respirable and Total Inhalable Dust*, MDHS 14, Health and Safety Executive, HMSO.

Figure 7.2b is taken from I. L. Marr and M. S. Cresser, *Environmental Chemical Analysis*, International Textbook Company, 1983.

Figure 7.4a is taken from S. E. Manahan, *Environmental Chemistry*, 4th Edn, Willard Grant, 1983.

Figure 8.2c is taken from R. Davis and M. Frearson, *Mass Spectrometry*, ACOL, Wiley, 1987.

1. Introduction

OVERVIEW

This part of the book discusses what is meant by the term 'environment', concerns over the effect of mankind on the environment and how this leads to a necessity for chemical monitoring.

1.1. THE ENVIRONMENT

Everyone nowadays seems to be concerned about **the environment**, but what do we mean by the term?

The place where we live or work?

The atmosphere which we breathe and water which we drink?

Unspoilt areas of the world which could soon be ruined?

Parts of the atmosphere which shield us from harmful radiation?

The environment must include all these areas and any other area which could affect the well-being of living organisms.

Concern must extend over any process which would affect this well-being, whether it is physical (e.g. global warming), chemical (e.g. ozone layer depletion) or biological (e.g. destruction of rain forests).

Anyone who has more than a passing interest in the environment has to learn and understand a very broad range of subjects. The purpose of

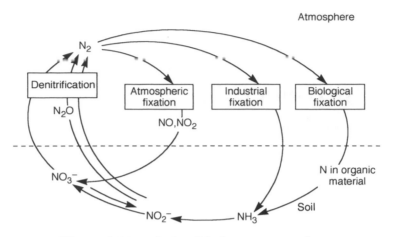

Figure 1.1a. *A simplified nitrogen cycle*

this introduction is first of all to show how analytical chemistry fits into this broad spectrum and later to demonstrate how it is an essential part of any scientific study of the environment and its problems. The book then goes on to discuss how analytical chemistry is applied to the three spheres of the environment, namely water, land and atmosphere.

To understand the environment, we must first realise that it is never static. Physical forces continuously change the surface of the earth through weather, the action of waves and natural phenomena such as volcanoes. At the same time they introduce gases, vapour and dust into the atmosphere. These can return to the land or sea a great distance from their source. Chemical reactions high up in the atmosphere continuously produce ozone, which protects us from harmful ultraviolet radiation from the sun. Living organisms also play a dynamic role through respiration, excretion and ultimately death and decay, recycling their constituent elements through the environment. This is illustrated by the well known nitrogen cycle (Fig. 1.1a) and similar cycles for all elements which are used by living organisms.

1.2. REASONS FOR CONCERN

The current interest in the environment stems from the concern that the natural processes are being disrupted by man to such an extent that the quality of life, or even life itself, is being threatened.

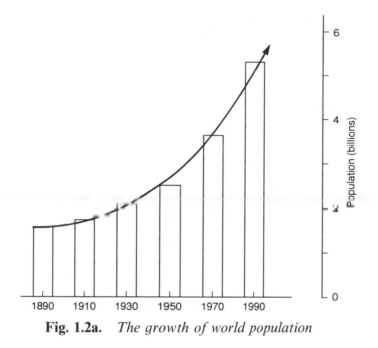

Fig. 1.2a. *The growth of world population*

Many indicators suggest that the world is at a crisis point, for instance the rapid population growth of the world as shown in Fig. 1.2a and the consequential growth in energy consumption (Fig. 1.2b), derived from

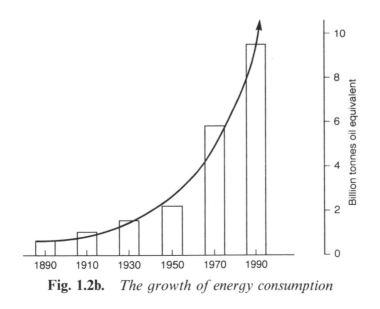

Fig. 1.2b. *The growth of energy consumption*

natural resources. Not only will the earth be depleted of its resources, with the inevitable environmental damage, but there will almost certainly be a parallel increase in waste produced and in pollution of the earth.

This concern has become heightened by greater awareness of problems than in previous ages, owing to greater ease of communication, which brings news from distant parts of the world. It seems ironic that the greater prosperity of the developed world, giving sufficient leisure time for concern over global problems, but also giving increased resource consumption, is currently a large contributing factor to the problems themselves.

1.3.   POLLUTION

All of us have concepts of what pollution is, but have you considered how it may be defined?

∏   What you would consider to be a definition of pollution?

The following definition is from the Organisation of Economic Co-operation and Development:

> Pollution means the introduction by man, directly or indirectly, of substances or energy into the environment resulting in deleterious effects of such a nature as to endanger human health, harm living resources or interfere with amenities or other legitimate use of the environment.

Before we concentrate on the chemical aspects of pollution, it is worth remembering that this is not the only form of pollution. Noise is an example of physical pollution. Simply adding water to a river at a different temperature to the ambient can affect life in the river. This is a form of thermal pollution. Pollution is, however, often associated with the introduction of chemical compounds into the environment. Popular opinion usually sees these as unnatural (and therefore harmful) substances. Perhaps one of the best known current examples is the concern over the emission of chlorofluorocarbons (CFCs). These have been used in aerosol sprays and other applications and are linked with the depletion of ozone in the stratosphere, which could lead to an increase in intensity of harmful ultraviolet radiation from the sun reaching the earth's surface.

More frequently problems occur by the release of substances into the environment which are naturally present, the problem arising simply from an increase in concentration above 'natural' levels.

Carbon dioxide is a natural component of the atmosphere produced by the respiration of living organisms. The problem of global warming is primarily associated with an increase in its concentration in the atmosphere as a result of fuel combustion. Increasing concentrations of a number of other naturally occurring gases, such as methane, add to the problem.

Nitrates occur naturally as part of the constant cycling of nitrogen in the environment (Fig. 1.1a). Over-use of fertilisers can, however, produce a build-up of nitrate in watercourses which leads first of all to excessive plant growth, but ultimately to the death of all living species in the water. The process is known eutrophication. Apart from nitrogen, all of the species in the nitrogen cycle have been shown to exhibit environmental problems if their concentration increases greatly above the 'natural' level in water or in the atmosphere. This is summarised in Fig. 1.3a.

N_2O contributes to the greenhouse effect and is a potential ozone depleter

NH_3 is highly poisonous in water to fish, particularly in its non-protonated form

NO_2^- is highly poisonous in water to animals

NO_3^- contributes to eutrophication (excessive plant growth) in watercourses and is associated with blue-baby syndrome in infants

Fig. 1.3a. *Examples of problems caused by excessive concentrations of nitrogen species*

Parameters Concerning Substances Undesirable in Excessive Amounts[a]

	Parameters	Expression of the results[a]	Guide level (GL)	Maximum admissible concentration (MAC)	Comments
20	Nitrates	NO_3 mg l^{-1}	25	50	
21	Nitrites	NO_2 mg l^{-1}		0.1	
22	Ammonium	NH_4 mg l^{-1}	0.05	0.5	
23	Kjeldahl nitrogen (excluding N in NO_2 and NO_3)	N mg l^{-1}		1	
24	(KMnO$_4$) oxidizability	O_2 mg l^{-1}	2	5	Measured when heated in acid medium
25	Total organic carbon (TOC)	C mg l^{-1}			The reason for any increase in the usual concentration must be investigated
26	Hydrogen sulphide	S μg l^{-1}		undetectable organoleptically	

No.	Parameter	Expression of the results			Observations
27	Substances extractable in chloroform	mg l⁻¹ dry residue — $mg\,l^{-1}$ dry residue	0.1		
28	Dissolved or emulsified hydrocarbons (after extraction by petroleum ether); mineral oils	$\mu g\,l^{-1}$		10	
29	Phenols (phenol index)	$C_6H_5OH\ \mu g\,l^{-1}$		0.5	Excluding natural phenols which do not react to chlorine
30	Boron	$B\ \mu g\,l^{-1}$	1000		
31	Surfactants (reacting with methylene blue)	$\mu g\,l^{-1}$ (lauryl sulphate)		200	

[a]Certain of these substances may even be toxic when present in very substantial quantities.

Fig. 1.3b. *Extract from European Community Council Directive 80/778/EEC relating to the quality of water intended for human consumption*

You should be able to think of many pollution examples of your own. Try grouping the problems into different categories, for instance whether the pollution is a global problem (e.g. ozone depletion) or a more local issue (e.g. waste dumping). When you read the next chapter, which deals with the transport of pollutants, you may find, however, that you change your mind about some of the problems. For instance, consider the increased lead concentrations in the environment, largely from the use of leaded petrol in cars, which can be detected hundreds of kilometres from likely sources. Lead pollution has been associated with the retardation of intellectual development in children, but is normally thought to be a highly localised problem.

∏   If a pollutant is discharged into the environment, what causes the effect on individual living organisms?
      — the total amount discharged?
      — its concentration in the environment?

It is the concentration which is of concern with respect to individual living organisms. This statement may seem surprising, but consider the following two facts:

(*a*) All compounds are toxic at high enough concentration. Even something apparently as innocuous as sodium chloride has adverse effects when present in high concentration. You cannot drink more than a small quantity of sea water without being made ill.

(*b*) Some metals, which are necessary for plant growth when found in small concentrations in the soil, would kill the plant life when found in higher concentrations on, let us say, a waste dump. These include elements such as chromium, cobalt and manganese, and are often known as 'essential' elements.

Of course, if we are considering the effect of a particular pollutant on the global environment, we would have to consider the total quantity emitted. Excessive amounts would ultimately increase the background concentration, as is the case with carbon dioxide emissions.

It would appear, then, that in order to limit the averse effect of a particular ion or compound, it is necessary to ensure that the concentration in water or in the atmosphere is maintained below a predetermined

level. As will be shown in the next section, the establishment of such
levels is fraught with difficulty. Nonetheless, much of the world's environmental legislation is drafted in terms of specifying maximum concentration of species in areas of the environment (Fig. 1.3b).

∏   What are the maximum concentrations of the substances listed
     in Fig. 1.3b that can be considered to be acceptable in drinking
     water?

These, of course, vary from substance to substance, but you should have
noted that most of the maximum admissible concentrations are expressed in mg l^{-1} units (sometimes called parts per million whereas
others are expressed as μg l^{-1} (or parts per billion).

SAQ 1.3a   How would you see the following situations as
contributing to pollution problems:
1. An increase in the developed world's population.
2. Volcanic emissions.
3. Production of methane by cows, as part of their
   natural digestion.
4. Excessive quantities of nitrate fertilisers used in
   farming.

1.4. THE NECESSITY OF ANALYSIS

If you were performing a simple pollution monitoring exercise, it is
evident that detailed analysis of pollution levels would be an essential
part, but let us now consider a complete control programme and look
in detail at what stages chemical analysis would be necessary.

∏   List what steps would be necessary for a national government
      or international agency to control a potential pollution problem,
      starting from the initial recognition.

      At what stages would chemical analysis be involved?

(a) Recognition of the Problem

This would appear to be an obvious statement until you consider how
recently many pollution problems have become recognised. The term
'acid rain' originally referred to localised effects of sulphur oxides
(SO_2, SO_3) produced from coal combustion and was introduced in the
19th century. Trans-national problems have, however, only been recognised in the last two decades. The contribution of other chemical
compounds, such as nitrogen oxides (NO, NO_2), to acid rain has only
been acknowledged in the last few years.

(b) Monitoring to Determine the Extent of the Problem

As we have already seen, this may either involve analysis of a compound
not naturally found in the environment, or determination of the increase
in concentration of a compound above the 'natural' level. The determination of 'natural' levels could involve a substantial monitoring exercise
since these levels may vary greatly with location and season. Mankind
has been introducing large quantities of some materials into the environment for many centuries, and it may even be a difficult task to assess
what an unpolluted environment is. For example, it has been discovered
that the highly toxic and potentially carcinogenic compounds commonly
referred to as 'dioxins,' which were originally assumed to be completely
anthropogenic (man-made), occur naturally at trace levels.

(c) Determination of Control Procedures

Possibilities include technological methods, such as the use of flue gas desulphurisation processes to control sulphur oxide emissions from power stations, and socially orientated methods, such as the promotion of the use of public rather than private transport to reduce vehicle emissions. Determination of the most appropriate method should involve testing the options with suitable analytical monitoring.

(d) Legislation to Ensure the Control Procedures are Implemented

Few pollution control methods are taken up without the backing of national or international legislation.

(e) Monitoring to Ensure the Problem has been Controlled

This also provides information for possible further improvements in legislation.

Have you noticed the cyclical nature of the process which includes monitoring to show that a problem exists, reduction of the problem by control procedures and monitoring to confirm that the problem has been reduced, with the final stage leading back to the start for improvement in the control procedures?

You should also have noticed that chemical analysis is a necessary component of almost all the stages!

SAQ 1.4a

Consider a factory producing a liquid discharge, consisting partly of side products of the process and partly of contaminants present in the starting materials. What analytical monitoring programme would be useful to assess and control the effluent?

SAQ 1.4a

Summary

This part of the book answers the question of what is meant by the terms 'environment' and 'pollution'. Pollutants are often materials which are naturally present in the environment, their adverse effects being caused by concentrations higher than those which would be expected from natural causes. A study of pollution would then involve a large amount of quantitative chemical analysis. Analytical chemistry is also involved in devising pollution control procedures, in drafting legislation and in monitoring the effect of any control procedure. In fact, analytical chemistry is a necessary component in almost all aspects of scientific investigations of the environment, the problems caused by mankind and their possible solution.

Objectives

You should now be able to:

● describe what is meant by the term 'environment';

● explain the reasons for concern over the current and future quality of the environment;

● demonstrate an appreciation of the diversity of pollution;

● evaluate the role of chemical analysis in dealing with these problems.

SAQs AND RESPONSES FOR PART ONE

SAQ 1.3a

> How would you see the following situations as contributing to pollution problems:
> 1. An increase in the developed world's population.
> 2. Volcanic emissions.
> 3. Production of methane by cows, as part of their natural digestion.
> 4. Excessive quantities of nitrate fertilisers used in farming.

Response

1. You could write many pages on the first question. The particular problem mentioned in Section 1.2 is the disproportionate proportion of the world's resources used in the developed world, together with a similar proportion of the waste produced and pollution of the earth.

2. Although volcanic emissions are natural phenomena, at intervals they put into the atmosphere large amounts of gases, vapour and dust, and could be considered a natural source of pollution.

3. The production of methane by cows is again a natural phenomenon. However, since the total cattle population on the earth is largely controlled by mankind, then so is the quantity of methane emitted to the atmosphere from this source.

4. This will directly lead to an increase in the concentration of nitrate above the naturally occurring levels in water supplies surrounding the farms. There may, however, be other consequences. Since all the species in the nitrogen cycle (Fig. 1.1a) are linked, changes in nitrate concentrations may lead to changes in concentrations of other species in the cycle, leading to further pollution problems (Fig. 1.3a).

These are all examples of current concern. As countries become developed, more of the earth's resources are used. The volcano Mount Etna is thought to be a significant contributor to atmospheric mercury concentrations in Europe. The rising atmospheric methane concentration has, in part, been attributed to increasing cattle populations. Concern has been expressed over the effect of increased nitrate fertiliser usage in the production of nitrous oxide, a greenhouse gas and potential ozone layer depleter.

SAQ 1.4a | Consider a factory producing a liquid discharge, consisting partly of side products of the process and partly of contaminants present in the starting materials. What analytical monitoring programme would be useful to assess and control the effluent?

Response

Analysis of the discharge before dispersal into the river will monitor the pollutant being discharged, but will only be related to the final concentration in the river when used in conjunction with river flow data.

Analysis of the river sufficiently far downstream from the discharge point to allow for dispersal will give a direct measurement of the concentration in the river, but there will be some uncertainty as to the source of the pollution.

Analysis of the pollutant in living organisms found in the river will give a direct indication of the environmental problem, but unless the organisms are sampled close to the discharge point, the analytical results would be difficult to relate to individual discharges.

The discharge composition was included to encourage you to extend your ideas to consider what extra analytical information would be useful. Quality control of the starting materials would reduce the concentration of the contaminants in the discharge, as would process control to minimise the manufacture of side products.

2. Transport of Pollutants in the Environment

OVERVIEW

This part of the book introduces you to how chemical compounds disperse, reconcentrate and degrade in the environment. From this we can start to select areas where high concentrations are likely to be found which will provide suitable sampling locations. Two classes of compound of major environmental concern are discussed in detail to illustrate the principles. These are high relative molecular mass neutral organic compounds, and metals.

2.1. INTRODUCTION

We have learnt how the environmental effects of compounds are dependent on their concentration and also that the environment is not static. Materials are constantly being transported between the three spheres of the environment—the atmosphere, the hydrosphere and the lithosphere (the earth's crust). At each stage of the transportation, the concentration of the compounds will be altered either by phase transfer, dilution or, surprisingly, reconcentration. Before discussing analytical methods we need to understand these processes so that we can:

— predict where large concentrations of the pollutant are likely to occur;

— assess the significance of measured concentrations of pollutant in different regions of the environment.

For this we need to discuss the chemical and physical properties of the pollutant. This will also help us to identify species which may be of particular concern and to understand why, of the many thousands of ions and compounds regularly discharged into the environment, particular concern often centres on just a few classes.

2.2. SOURCES, DISPERSAL, RECONCENTRATION AND DEGRADATION

Virtually every form of human activity is a potential source of pollution. The popular concept of industrial discharge being the primary source of all pollution is misguided. It is just one example of a *point* source, i.e. a discharge which can be readily identified and located. Discharges from sewage work provide a second example. In some areas these are the major source of aquatic pollution.

Sometimes, however, it is not possible to identify the precise discharge point. This can occur where the pollution originates from land masses. Examples include the run-off of nitrate salts into watercourses after fertiliser application and the emission of methane from landfill sites into the atmosphere. These are examples of *diffuse* sources.

Both water and the atmosphere are major routes for the dispersal of compounds. What comes as a surprise are the pathways by which some of the compounds disperse. It is very easy, for instance, for solid particulate material to be dispersed long distances via the atmosphere. There is, for example, an approximately equal quantity of lead entering the North Sea off the coast of Britain from atmospheric particulates as from rivers or the dumping of solid waste. To illustrate this a typical transport scheme for a metal is shown in Fig. 2.2a.

Equally surprising are the dispersal routes of 'water-insoluble' solid organic compounds. No material is completely insoluble in water. For instance, the solubility in water of the petroleum component isooctane is as high as 2.4 mg l^{-1}. Watercourses provide a significant dispersal route for these compounds.

The significant vapour pressure of organic solids is also often forgotten. Consider how readily a solid organic compound such as naphthalene,

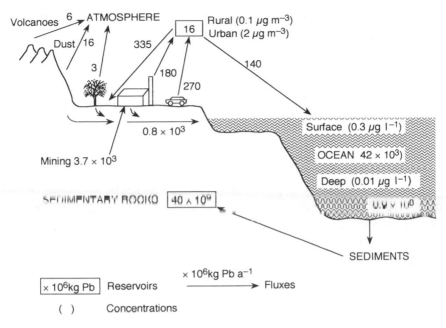

Fig. 2.2a. *Transport of lead in the environment*

as used in mothballs, volatilises. In these cases transportation through the atmosphere is partly in the solid phase and partly in the vapour phase. If you wish to monitor the concentration of these materials in the atmosphere, you have to analyse not only the gaseous fraction, but also the suspended particulate material.

The atmosphere also provides a dispersal route for volatile organic compounds. Hydrocarbons will be quickly degraded but will contribute to localised pollution in the form of photochemical smog. If the compound is stable, or is only slowly degraded, in the lower atmosphere, as is the case with many chlorine- or bromine-containing compounds, some may eventually reach the stratosphere (the portion of atmosphere at an altitude of 10–50 km). Decomposition promoted by the intensity of low-wavelength radiation at this altitude initiates a series of chemical reactions, depleting the protective layer of ozone.

Distances which are travelled by pollutants in the atmosphere may be hundreds or thousands of kilometres. The movement of sulphur oxides has been studied over distances covering the whole of Europe and when the Mount St Helens volcano erupted in the USA, the particulate material discharged resulted in the production of vivid sunsets even several thousand kilometres away.

Dispersal of a pollutant in water or in the atmosphere will inevitably lead to a dilution of the pollutant. As we have seen that the effect of a chemical compound in the environment can be related directly to its concentration, you may think that the dispersal process will simply spread out the pollutant such that it could have little effect away from the source. This would especially be the case when we consider that most forms of pollution are eventually broken down by microbial attack, photochemical or other degradation and so there would be little chance of the concentration building up to toxic levels. Indeed, the phrase 'Dilution is the solution to pollution' was often heard in the early days of environmental concern.

∏   What factors do you think this statement does not take into account?

(*a*) The possibility that some pollutants can reconcentrate at particular locations or within organisms remote from the original source.

(*b*) The non-degradation or slow degradation of some pollutants so that there is a gradual concentration build-up in the environment at large.

(*c*) Contamination of large areas before sufficient dilution has taken place.

Examples may be given for all these cases:

(*a*) Toxic metals, such as cadmium, may be found in the organs of shellfish in concentrations up to 2 million times greater than in the surrounding water (Fig. 2.2b).

(*b*) The major constituent of the pesticide DDT (*p,p'*-dichlorodiphenyl-trichloroethane) is now a universal contaminant owing to its widespread use over several decades and its slow degradation. There is little organic material on the earth which does not contain traces of this at the $ng\ kg^{-1}$ level or greater concentration.

Metal	Relative concentration in shellfish (water = 1)
Cadmium	2 260 000
Chromium	200 000
Iron	291 500
Lead	291 500
Manganese	55 500
Molybdenum	90
Nickel	12 000

Fig. 2.2b. *Examples of metal enrichment in shellfish relative to the surrounding water*

(c) Dilution does not take into account localised pollution effects which may occur around discharge pipes or chimneys before dispersion occurs.

The effects of pollution have also in the past often been underestimated. The discharge of sulphur dioxide in gases from tall chimneys was, until recently, seen as an adequate method for its dispersal. The potential problem of 'acid rain' was not considered.

The following sections will discuss two major categories of pollutants which have caused environmental concern owing to their ability to reconcentrate (accumulate) in specific areas and within living organisms. They provide good examples of how a knowledge of the transport of pollutants can be used to determine suitable sampling locations where high concentrations may be expected.

SAQ 2.2a | What general physical and chemical properties would you expect in a compound which has become a global pollution problem?

SAQ 2.2a

2.3. TRANSPORT AND RECONCENTRATION OF NEUTRAL ORGANIC COMPOUNDS

Compounds in this category which readily reconcentrate and are of global concern are usually of low volatility and high relative molecular mass ($M_r > 200$). They often contain chlorine atoms within the molecule. Typical compounds are shown in Fig. 2.3a.

Compounds of lower relative molecular mass may produce severe local atmospheric problems. Hydrocarbon emissions from automobiles are currently of concern owing to their contribution to the photochemical smog which affects large cities throughout the world. These effects occur where the climate and geographical conditions permit high atmospheric concentrations to build up with little dispersal. However, unless the compounds are particularly stable to decomposition within the atmosphere (as is the case with chlorofluorocarbons), or are discharged in such great quantities that they can build up globally (as is the case with methane), they will remain local, rather than global, pollutants.

We shall now discuss the mechanisms by which organic compounds can reconcentrate to high concentrations within organisms, and we shall discover one of the reasons why it is the compounds of higher relative molecular mass that are of greatest concern.

1. Organochlorine pesticides

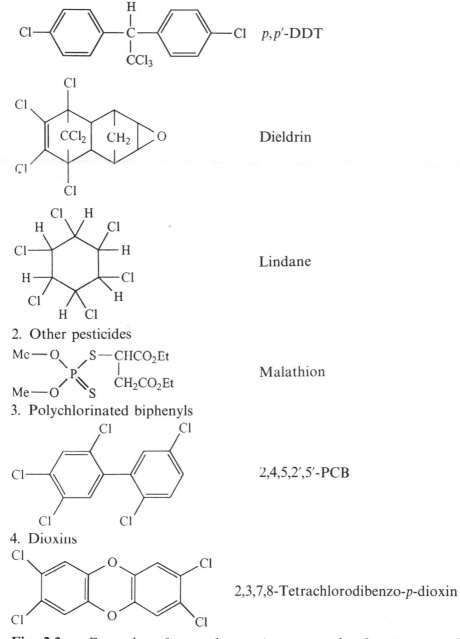

p,p'-DDT

Dieldrin

Lindane

2. Other pesticides

Malathion

3. Polychlorinated biphenyls

2,4,5,2',5'-PCB

4. Dioxins

2,3,7,8-Tetrachlorodibenzo-*p*-dioxin

Fig. 2.3a. *Examples of neutral organic compounds of environmental concern*

2.3.1. Bioconcentration

Unless organic compounds contain polar groups such as OH or NH_2, or are ionic, they will have low solubility in water. Within related groups of compounds the solubility decreases with increasing molecular mass. As the solubility in water decreases the solubility in organic solvents increases (Fig. 2.3b).

This increase in solubility is equally true if we consider solubility in fatty tissues in fish and aquatic mammals rather than solubility in laboratory solvents. Any dissolved organic material will readily transfer into fatty tissue, particularly that found in organs in closest contact with aqueous fluids, e.g. kidneys.

∏   What rule can you deduce concerning the solubility of a compound in water and its ability to accumulate in organisms?

We arrive at a very unexpected general rule that the lower the solubility of an organic compound in water, the greater is its ability to accumulate in fatty tissues and the greater is the potential for toxic effects. Also,

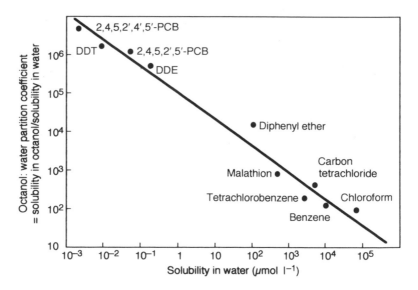

Fig. 2.3b. *Partition coefficients versus aqueous solubilities of environmentally significant organic chemicals*

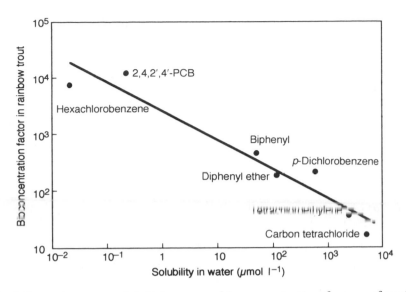

Fig. 2.3c. *Aqueous solubilities versus bioconcentration factors of environmentally significant organic chemicals in rainbow trout*

because the solubility in water decreases with increasing molecular mass for related groups of compounds, we could also deduce that higher molecular mass compounds may pose greater aquatic environmental problems than compounds of lower molecular mass.

The rule is illustrated by Fig. 2.3c, where the ability to accumulate in an organism is measured by the bioconcentration factor, where

$$\text{bioconcentration factor} = \frac{\text{concentration of a compound in an organism}}{\text{concentration in surrounding water}}$$

2.3.2. Accumulation in Sediments

This is also related to the low solubility of high molecular mass organic compounds in water, together with the hydrophobicity of organic compounds not containing polar groups. Undissolved or precipitated organic material in water will adhere to any available solid; the larger the solid surface area, the greater will be its ability to adsorb the compound. Suitable material is found in sediments. This is particularly true within estuaries where there are often discharges from major industries and fine sediment is in abundance. It is often the case (as may be expected from surface area considerations) that the smaller the particle size then the greater the accumulation of organic compounds in the sediment.

These organics may then be ingested by organisms which feed by filtration of sediments (e.g. mussels, scallops) or, if the solid is sufficiently fine to be held in suspension, by bottom-dwelling fish.

2.3.3. Biomagnification

Animals obtain their food by feeding on other plants or animals. Food chains can be built up where one species is dependent for survival on the consumption of the previous species.

If a pollutant is present in the first organism, then as we proceed down the food chain there will be an increase in concentration in each subsequent species. This is illustrated in Fig. 2.3d.

Although the concept of food chains is much simplified from what happens in nature (few species have just one source of food), it does provide an explanation of why the greatest concentration of pollutants is found in birds of prey at the end of the food chain, rather than in organisms in closest contact with the pollutant when originally dispersed.

2.3.4. Degradation

Even if a compound has a tendency to transfer into organisms by the routes described, it will not build up in concentration within the organism if it is rapidly metabolised. Compounds will break down until a molecule is produced with sufficient water stability to be excreted. The solubility may be due either to polar groups being attached to the molecule or to its low relative molecular mass.

The rate of metabolism is highly dependent on the structure of the molecule. One of the reasons why so many organic compounds of

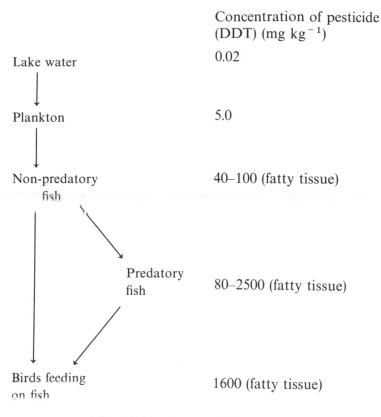

	Concentration of pesticide (DDT) (mg kg⁻¹)
Lake water	0.02
Plankton	5.0
Non-predatory fish	40–100 (fatty tissue)
Predatory fish	80–2500 (fatty tissue)
Birds feeding on fish	1600 (fatty tissue)

Fig. 2.3d. *A typical food chain*

environmental concern contain chlorine atoms is the slow metabolism of many of these compounds.

If we take as an example *p,p'*-DDT, the metabolism of this compound occurs in two stages, as shown in Fig. 2.3e.

The first stage is rapid and normally takes only a few days for completion but the second stage is extremely slow, often taking many months in some species. It is, in fact, the first degradation product which is often the predominant species in environmental samples.

A minor component of commercial DDT is the *o,p'*-isomer. This is metabolised rapidly by the reaction shown in Fig. 2.3f and so does not accumulate significantly in organisms.

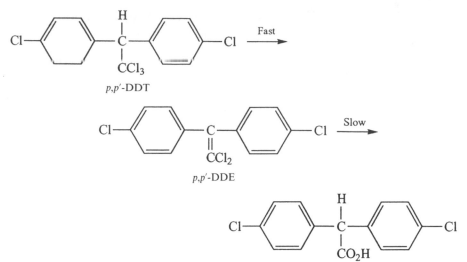

Fig. 2.3e. *Metabolism of* p,p'-*DDT*

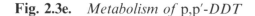

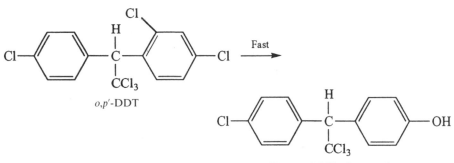

Fig. 2.3f. *Metabolism of* o,p'-*DDT*

SAQ 2.3a

Consider a pesticide such as DDT being sprayed on to a field from an aeroplane. Sketch routes by which the pesticide may disperse from the area of application.

SAQ 2.3a

2.4. TRANSPORT AND RECONCENTRATION OF METAL IONS

We were able to discuss the movement of neutral organic compounds in simple terms because often very little chemical change occurs to the compounds during transportation through the environment and the initial degradation products frequently have similar physical and chemical properties to the parent compound. Unfortunately, this is not the case with many of the metals of environmental concern. Their reaction products often have vastly different chemical and physical properties.

The metals which are of most environmental concern are first transition series metals and post-transition metals (Fig. 2.4a), many of which are in widespread use in industry. Often the non-specific term 'heavy metals' is used for three of the metals, lead, cadmium and mercury. These have large bioconcentration factors in marine organisms (look at the values for lead and cadmium in Fig. 2.2b), are highly toxic and, unlike many of the transition elements, have no known natural biological function.

The following paragraphs introduce you to the chemical principles which can govern the transportation of metals in the aquatic environment and give indications as to where high concentrations may be found.

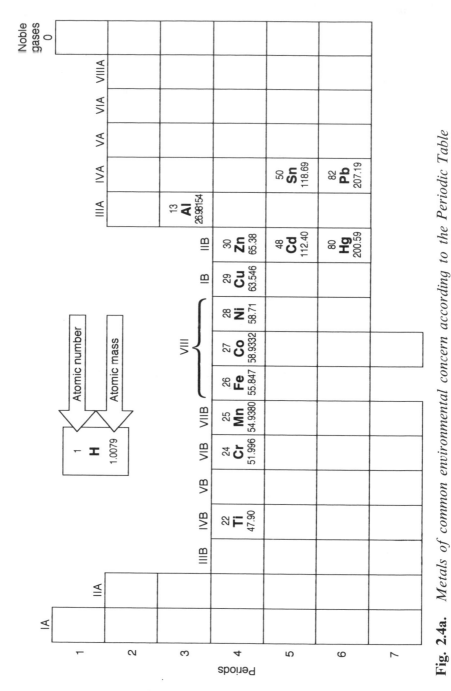

Fig. 2.4a. *Metals of common environmental concern according to the Periodic Table*

2.4.1. Solubilisation

Metals entering the environment are often in an insoluble form in industrial waste, in discarded manufactured products or as part of naturally occurring mineral deposits. Deposition from the atmosphere is often in the form of insoluble salts. However, the solubility of metals increases with decrease in pH. Some of the problems of 'acid rain' in causing the death of fish have been attributed to the leaching of toxic metals from the soil, in addition to the direct effect of pH on the fish. The use of lead pipes for domestic water supplies is more problematic in areas of soft, acidic water than where the water is hard and slightly alkaline.

Solubilisation is often aided by the formation of complexes with organic material. These may be anthropogenic (e.g. complexing agents in soap powders), but may also occur naturally. Humic and fulvic acids produced by decay of organic material can help solubilise metals.

2.4.2. Deposition in Sediments

This can occur when there is an increase in pH. The pH at which this occurs may vary from metal to metal but under sufficiently alkaline conditions all transition metals will precipitate. Deposition of relatively high concentration metals may result in traces of other metal ions also being deposited. This is known as coprecipitation. Metal ions may also interact with sediments by a number of mechanisms including:

adsorption;

ion exchange (clay minerals are natural ion exchangers);

complex formation within the sediment.

A change in the oxidising or reducing nature of the water (i.e. the redox potential) may lead to either solubilisation or deposition of metal ions.

Most transition metal ions can exist in a number different oxidation states in solution (e.g. iron can exist as Fe^{2+} and Fe^{3+}). Iron in solution under slightly acidic conditions is predominantly Fe^{2+}. Under alkaline oxidising conditions the iron is oxidised and precipitates as $Fe(OH)_3$.

Under reducing conditions all sulphur-containing ions (e.g. SO_4^{2-}) are reduced to S^{2-}, and this may lead to the deposition of metals such as lead and cadmium as their insoluble sulphides.

2.4.3. Uptake by Organisms

From the above considerations, an obvious route into the food chain is from sediments via filter feeders. Many metals are retained in the organism as a simple ion. Others, particularly cadmium and mercury, can be converted into covalent organometallic compounds. These will behave in a similar fashion to covalent organic compounds described previously and will preferentially accumulate in fatty tissues. The distribution of the metal within an organism is thus very dependent on the individual metal and its detailed chemistry. Compare the distribution of lead and cadmium in shellfish in Fig. 2.4b.

Sample	Percent of whole animal	Concentration (mg kg^{-1})	
		Pb	Cd
Scallop			
Gills	10	52	<20
Muscle	24	<5	<20
Fatty Tissue	17	8	2000
Intestine	1	28	<20
Kidney	1	137	<20
Gonads	20	78	<20
Sediment	—	<5	<20
Sea water (mg l^{-1})	—	3	0.11

Fig. 2.4b. *Concentrations of trace elements in individual organs of shellfish*

∏ One metal which is of current environmental concern cannot be described either as a transition metal or as a 'heavy' metal. What is this metal?

Aluminium. It is found in great abundance in the aluminosilicate structures of clays, but is usually fixed in this insoluble form. When the acidity increases sufficiently this solubilises the aluminium.

SAQ 2.4a

> Compare the routes by which high molecular mass organic compounds and toxic metals may disperse and reconcentrate in the environment and in organisms.

2.5. WHAT IS A SAFE LEVEL?

We have now discussed many of the concepts needed to determine the movement of pollutants in the environment and, if degradation of the compound is slow, how reconcentration may occur.

Interpretation of the analytical data needs to be based on the relationship between the analytical concentration and the effect on organisms. This correlation may not be as easy to determine as first may be thought.

Toxicological testing has been performed on many (but by no means all) compounds which produce major environmental problems. The testing is generally under short-term, high-exposure ('acute' exposure) conditions. This may take the form of determining the dose or concentration likely to cause death to a percentage of test organisms. The LD_{50} test, for example, determines the lethal dose required for the death of 50% of the sample organisms. This testing is, however, not generally relevant to environment problems, where it is much more likely that the exposure is over a long term in small doses or low concentrations ('chronic' exposure). The effect may be non-lethal, such as a decrease in the rate of growth or an increase in proportion of mutations in the offspring, but over several generations still leads to a decrease in the population of the species. Monitoring of chronic effects may not be easy outside the laboratory, and may be complicated by the presence of other pollutants, or other uncontrollable effects (e.g. climate). One of the reasons why the environmental problems of p,p'-DDT are often discussed is that its initial release in the early 1940s was into an environment largely free from similar pollutants. Possible effects could be readily correlated with analytical concentrations. This is not as easy nowadays as any compound under investigation will invariably be present in organisms as part of a cocktail with other compounds.

This leads us to the next problem that the effect of two or more pollutants together may be greater (synergism) or less (antagonism) than that predicted from the two compounds individually. For instance, the effect of sulphur dioxide and dust particles in some forms of smog is much greater than the separate effects of the two components. The toxicity of ammonia in water decreases with decrease in pH (i.e. with an increase in the hydrogen ion concentration). The ammonium ion, which is the predominant species under acidic conditions, is less toxic than the non-protonated molecule predominating under alkaline conditions.

The consequence of this for interpretation of analytical data is that information on the concentration on secondary components is often as important as the major analysis. This complicates the analytical task significantly.

2.6. WHERE DO WE TAKE SAMPLES?

Using the arguments in the previous sections, you should be able to predict routes by which a particular compound may be transported through the environment. We could start to predict which types of organisms would be most affected. This is a necessary preliminary step for any new monitoring programme in order to maintain the sampling within practicable limits. Even so, the analytical task could still be enormous. When the programme has become established the use of *critical paths* and *critical groups* can reduce the task. The critical path is the route by which the greatest concentration of the pollutant occurs and the critical group is the group of organisms (or people!) most at risk at the end of the critical path. If the concentration of the compound in samples taken from the critical group is within the permitted range, then it follows that concentrations will be no higher in other groups. Monitoring can be largely directed towards the assumed critical path and group but more widespread monitoring should continue—you my be wrong in your choice of path, or the conditions for which you deduced the path may change. The continuing programme should check these assumptions.

SAQ 2.6a

> Consider discharges of aqueous waste into a semi-enclosed sea area in which there is a thriving inshore fishing industry. If the waste consists of low-concentration transition and actinide metal salts, what would be the likely critical path and critical group of people?

2.7. GENERAL APPROACH TO ANALYSIS

We have seen how many ions and compounds can build up in concentration in organisms even when the background concentrations are in the $\mu g\,l^{-1}$ range. In some instances where the compound is highly toxic, resistant to biodegradation and bioaccumulates very readily, concern is expressed even when the concentrations approach the limit of experimental detection. This is the case with the dioxin group of compounds, which are routinely monitored at $ng\,l^{-1}$ concentrations.

At the other end of the concentration scale, monitoring is often required in water for components which may be present in tens or hundreds of $mg\,l^{-1}$. In these cases, the analysis may not necessarily be specific to individual ions or compounds as the measurements are often concerned with bulk properties of the water (e.g. acidity, water hardness). These are often known as 'water quality' parameters.

∏   From your knowledge of analytical techniques, briefly list the types of method which may find use in environmental analysis for organic compounds and metals. How would these fit into a more complete analytical scheme for a typical low concentration environmental component?

The broadest categories which you may have listed are probably:

(*a*) classical methods of analysis, i.e.

  volumetric methods;

  gravimetric methods;

(*b*) instrumental methods.

You may then have subdivided the instrumental methods, but we will start with these broad divisions.

Volumetric analyses (titrations) are rapid and accurate, use simple and inexpensive apparatus and can be used for direct measurements of the bulk properties. Water hardness, for instance, can be measured by a single titration regardless of the nature of the ions producing the effect. They are, however, of limited use for concentrations below the $mg\,l^{-1}$ level, and (although automation is possible) can be labour intensive.

Gravimetric techniques can be of extremely high accuracy, but very prone to interference from other species. A high degree of skill is necessary for accurate analyses. They tend to be slow techniques owing to the time taken for precipitation, filtration and drying. Gravimetric methods are used as reference methods to check the accuracy of other techniques.

Instrumental methods are usually more suited to low concentrations. The linear operating range of instrumentation is often at the $mg\,l^{-1}$ level, frequently corresponding very closely to environmental concentrations. The analysis of the sample is generally rapid and can easily be automated. You should be aware, however, that sample preparation and instrument calibration can often be more time consuming. Accuracy is lower than for the classical techniques but is usually sufficient for most applications.

Most of the instrumental methods we shall be discussing fit into one of the following categories:

chromatographic methods;

spectrometric methods;

electrochemical methods.

As already mentioned, the methods may be sufficiently sensitive for many analyses, often with little sample preparation. Preconcentration of the sample may be used to decrease the lower detection limits of the techniques. In addition, a pre-analytical separation stage may be included to remove interfering components. We can then construct a typical analytical scheme which will cover many of the methods discussed in later sections:

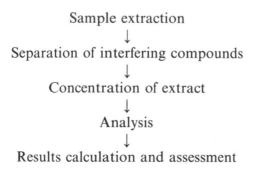

Sample extraction
↓
Separation of interfering compounds
↓
Concentration of extract
↓
Analysis
↓
Results calculation and assessment

Summary

Pollutants travel through the environment by routes which can be predicted from their chemical and physical properties. Pollutants of major concern include high relative molecular mass neutral organic compounds and metals. They are capable of reconcentrating in certain areas and within organisms and it is in these areas that they have their greatest effect. An understanding of such routes is needed for the correct choice of sampling positions for subsequent analytical determinations.

Objectives

You should now be able to:

• predict the possible movements of a pollutant in the environment;

• suggest sampling locations where high molecular mass organic compounds and metals may accumulate;

• define what is meant by the terms critical path and critical group;

• understand the range of methods needed for subsequent chemical analysis.

SAQs AND RESPONSES FOR PART TWO

SAQ 2.2a   What general physical and chemical properties would you expect in a compound which has become a global pollution problem?

Response

The first prerequisite is that the compound must be in a form which allows it to become widely dispersed. Dispersal may be via the atmosphere or the hydrosphere. The properties which would affect this dispersal include volatility, solubility in water and, if the compound is a solid, its particle size.

The compound must have a high resistance to degradation within the atmosphere and hydrosphere and to metabolism (chemical reaction) within organisms. If the compound rapidly degraded, there would be little possibility of toxic concentrations building up. Beware, however, of the possibility that the degradation products themselves might pose similar or worse environmental problems.

If the pollution problem is related directly to the effect of the material on living organisms, rather than on physical structures, then the compound must have an ability to reconcentrate within organisms. This will be described in detail in the following sections.

Finally, if the compound is seen as a pollution problem, it must have some deleterious effect. There are several classes of compound (e.g. phthalate esters used as plasticisers) which fulfil all the above prerequisites, including being dispersed globally, yet, according to our current state of knowledge, do not present a major environmental problem.

SAQ 2.3a

> Consider a pesticide such as DDT being sprayed on to a field from an aeroplane. Sketch routes by which the pesticide may disperse from the area of application.

Response

The dispersal of a pesticide from its area of application is shown in Fig. 2.3g.

Fig. 2.3g.   *Dispersal of a pesticide from its area of application*

SAQ 2.4a Compare the routes by which high molecular mass organic compounds and toxic metals may disperse and reconcentrate in the environment and in organisms.

Response

Both types of pollutant may be dispersed through the atmosphere and also in watercourses. Atmospheric dispersal of metals is usually as particulates (in the form of metal salts), whereas organics are found both in the particulate form and in the vapour state.

Deposition may occur on to land or into watercourses.

The solubility of the organic compounds in water is often low, but this can lead to a large bioconcentration in organisms living in the water. The solubility of metals is very dependent on the chemical compounds involved, but for most species will increase with a decrease in pH. There is normally no pH effect on the solubility of neutral organic compounds. The bioconcentration of metals is very dependent on the element being considered, but for some metals such as cadmium can be extremely high (Fig. 2.2b).

Both types of pollutant will concentrate in sediments. The detailed mechanisms for this are different in the two cases. Reducing conditions would, for instance, increase the deposition of lead, whereas they would have no effect on neutral organic compounds.

Entry into the food chain in an aquatic environment is in both cases by bottom-dwelling fish and filter-feeders. Once again the detailed mechanisms are different in the two cases and this may lead to concentration in different organs in the body.

If you had said there were some similarities in the dispersal and reconcentration you would have been correct, but there are many differences in the detailed mechanism.

SAQ 2.6a

> Consider discharges of aqueous waste into a semi-enclosed sea area in which there is a thriving inshore fishing industry. If the waste consists of low-concentration transition and actinide metal salts, what would be the likely critical path and critical group of people?

Response

The metal ions would be likely to concentrate in the sediment on the sea bed and be ingested by filter-feeders. A likely critical group would be the local people who consume large quantities of the sea food, perhaps the families of the fishermen themselves.

A second likely path could exist. Some of the sediment may be washed up on the shore. This may dry out and be blown into the atmosphere, or it may be propelled into the atmosphere in sea spray.

The critical group could then be people who spend a large proportion of their time on or near the seashore, the metal ions entering their bodies through inhalation.

3. Water Analysis— Major Constituents

OVERVIEW

This part of the book introduces you to the chemical composition of water as found in different parts of the environment. The analytical techniques which are used to determine the major constituents both in the laboratory and in the field are then described. The importance of correct sampling procedures is highlighted.

3.1. INTRODUCTION

Water is vital for life. Not only do we need water to drink and to wash but it is also important for many of the pleasant recreational aspects of life.

∏   List uses which we make of water.

This should include the following, but you may have thought of some extra ones of your own:

domestic water supply;
industrial water supply;
effluent disposal;
fishing;
irrigation;
navigation;
power production;
recreation, e.g. sailing, swimming.

Each different use has its own requirements for the composition and purity of water and each body of water to be used will need analysis on a regular basis to confirm its suitability. The types of analysis could vary from simple field testing for a single analyte to laboratory-based, multi-component instrumental analysis.

Water is found naturally in many different forms. In the liquid state it is found in rivers, lakes and ground water (water held in rock formations) and also as sea water and rain. As a solid it is found as ice and snow. Water in the vapour state is found in the atmosphere. You will certainly be familiar with the fact that sea water contains large quantities of dissolved material in the form of inorganic salts, but it may come as a surprise that nowhere in the environment can you consider water to be chemically pure. Even the purest snow contains components other than water.

∏   Write down some of the constituents which you consider might be found in natural river water.

Ions derived from commonly occurring inorganic salts, e.g. sodium, calcium, chloride, sulphate ions.

Smaller quantities of ions (e.g. transition metal ions) derived from less common inorganic salts, perhaps through leaching from mineral deposits.

Insoluble solid material, either from decaying plant material, or inorganic particles from sediment and rock weathering.

Soluble or colloidal compounds derived from decomposition of plant material.

Dissolved gases.

You will probably have written down most these. The category many forget to include are the dissolved gases. This, of course, includes oxygen, which is so vital in supporting aquatic life.

Dissolved gases occur through contact with the atmosphere and through respiration and photosynthesis. A fast-flowing turbulent river will usually be saturated in atmospheric gases. Respiration of aquatic animals will consume oxygen and produce carbon dioxide. Photosynthesis by plants reverses this.

Respiration produces energy from foodstuffs using oxygen:

$$C_6H_{12}O_6 + 6O_2 \longrightarrow 6CO_2 + 6H_2O + \text{energy}$$
glucose

Photosynthesis produces organic compounds using sunlight:

$$6CO_2 + 6H_2O + h\nu \quad \longrightarrow \quad C_6H_{12}O_6 + 6O_2$$

Oxygen levels in water are depleted by slow oxidation of organic and, in some cases, inorganic material. The presence of large quantities of oxidisable organic material (e.g. from sewage effluents) is often the most serious form of pollution in watercourses.

Ions commonly found in the $mg\,l^{-1}$ concentration range are the following:

Concentration range in natural river water ($mg\,l^{-1}$)	Cations	Anions
0–100	Ca^{2+}, Na^+	Cl^-, SO_4^{2-}, HCO_3^-
0 25	Mg^{2+}, K^+	NO_3^-
0–1	Other metal ions	PO_4^{3-}, NO_2^-

Others (e.g. fluoride ions) may occur depending on the mineral deposits in the locality.

Your list should also have included the compounds derived from decomposition of plant material. Did you include inorganic in addition to organic products? Do not forget ammonia, which can occur in water in the range $0–2\,mg\,l^{-1}$. Concentrations never usually increase to greater than these values as ammonia is rapidly oxidised to nitrate. It has significant toxicity to fish, particularly when it is present as the neutral molecule, rather than when protonated to form the ammonium ion.

Now look at Fig. 3.1a, showing typical comparative analyses for rain water, river water and sea water.

You will find similar ions in all three, the only difference being the concentration range. Sea water contains the common ions at the $g\,l^{-1}$

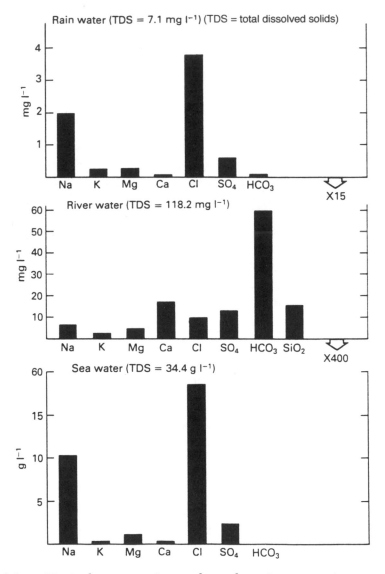

Fig. 3.1a. *Typical comparative analyses for rain water, river water and sea water. Note the different scales for each histogram*

level whereas for river and rain water the range is the mg l^{-1} level. All are easily measurable with modern instrumentation.

The situation would be slightly different if we tabulated the less common species. The range of ions (particularly metal ions) would be limited in river water by the chemical composition of the rocks over which it was flowing. On the other hand, sea water contains trace quantities of virtually every element, the highest concentrations being found close to the surface and in coastal areas. It is a very complicated analytical matrix indeed.

Have you noticed that, although the absolute concentrations within rain water and sea water are very different the relative concentrations are often very similar, giving a clue as to the origin of these ions?

∏   Estimate the sodium/potassium and potassium/calcium concentration ratios in rain, river and sea waters

Rain, especially when falling close to the sea, has sea water as a major constituent and can often be regarded as diluted sea water.

Detailed comparison of the concentrations of ions in a large number of rivers with those in sea water (which appears depleted in a number of elements, including calcium) is one of the methods of studying the complexities of marine chemistry. Unfortunately, further discussion of this is outside the scope of this book.

Water authorities often feel it necessary to analyse a river at many locations along its course. This is because the composition of water is never static. It changes by interaction with the atmosphere and crust, and by chemical and biological processes occurring within the water. This does not include the possibility of extra material being added in the form of pollution.

Let us consider a river flowing from its source to the sea. Even at its source water will contain dissolved salts from the passage of water through the earth to form the river. Some of the natural processes which will affect the composition are listed below:

(i) Weathering of rocks.

This will produce an increase in inorganic salt content. The composition may also be affected by interaction with material on the river bed. Clays, often found on river beds, are natural ion-exchange materials.

(ii) Sedimentation of suspended material.

As the river progresses downstream it will generally become less turbulent and so less capable of supporting suspended material.

(iii) Effect of aquatic life.

Absorption and emission of oxygen and carbon dioxide have already been mentioned. Living plants will also absorb nutrients (including nitrate and phosphate) necessary for growth.

The death and decay of organisms will release ions and also produce quantities of suspended material. This will slowly decompose into simpler chemical compounds. If the process proceeded to completion the final products would be carbon dioxide and water.

Dense beds of vegetation can also very effectively filter out suspended solids.

(iv) Volatilisation and evaporation.

Low relative molecular mass organic compounds tend to have a high vapour pressure and will be readily lost from water. A significant percentage of the water itself in the river can be lost through evaporation (the rate depending on the ambient temperature) and this will have the effect of increasing the concentration of all dissolved material in the river.

These are illustrated in Fig. 3.1b.

Fig. 3.1b. *Natural processes affecting river composition*

Even if one disregards the introduction of new compounds by pollution, water contains a large number of components. In fact, if one starts considering components which may be found at trace levels (less than $mg\,l^{-1}$) the task would be almost impossible as new components are constantly being identified in natural waters. Thankfully it is very rare that all the components would need to be analysed.

This section includes methods for the analysis of major components of water which may be routinely undertaken by water authorities. Even so, it would be unusual for all of the methods to be used on one sample. Water authorities or others undertaking the analyses will in general have a reasonable idea of what species to expect in the water. Unless there is a specific reason for more complete analysis, the analytical scheme will usually be restricted to components which are likely to cause environmental problems or exceed prescribed limits.

The analytical process has to start with sampling and sample storage, since changes in the composition of water do not stop once you have taken the sample. Precautions always have to be taken to make sure that the water reaching the laboratory has the same composition as it did when you took the sample.

SAQ 3.1a | Using the list you produced of likely chemical species in a river, decide which would be likely to increase or decrease downstream from a sampling point close to the source.

3.2. SAMPLING

The first task in any analysis is sampling, yet discussion of it is often neglected. The importance of correct procedures cannot be over-estimated as no matter how sophisticated the analytical equipment in the laboratory, it will only analyse the sample that is brought into the laboratory. The phrase often used when inaccurate data are sent for computer analysis, 'rubbish in ... rubbish out,' is just as applicable to chemical analysis. The sample (which often is only 250 or 500 ml) must be representative of the whole body of water requiring analysis. The sample must also be kept in such a manner that the concentration of the species to be analysed is unchanged during transportation and possible storage. This can be a major task when you consider the low concentrations often expected for pollutants and that the pollutants are frequently highly reactive or volatile.

Let us consider sampling a river. Water authorities will often have fixed protocols, but let us develop a scheme starting afresh.

(i) Before starting, decide on what analyses are required. The analytical techniques to be used will affect the sample size taken, the type of sample bottle and also the method of storage. It will be too late to alter these by the time you get back to the laboratory.

(ii) Decide on a sampling programme. We have already discussed how the composition of natural water is always changing. Sometimes the variation in composition may be periodic:

Seasonal—the concentration is affected by natural growth processes.

Weekly—a pollutant may only be emitted from a factory during the working week.

Daily—the concentration of some components may be changed owing to biological processes needing the presence of sunlight.

You may wish to monitor these regular fluctuations but you may be more concerned with the longer term variation of concentrations. Your sampling programme, the number of samples and the timing of the sampling will be affected. If you are interested in long-term variations it may be beneficial to take samples at the same stage of each periodic cycle, whereas for short-term variations you would take several samples each cycle.

∏   What regular variations would you expect in concentrations of the following:

dissolved oxygen;

nitrate.

Oxygen is produced by photosynthesis in the daytime but consumed by respiration or by oxidation of organic material continuously. (There will be a continuous but slow replenishment from the atmosphere.) A decrease in oxygen concentration during the night would be expected.

Variation of nitrate would be more complex. It is a nutrient necessary for growth and so if there were no additional inputs it would decrease in the spring growing season and increase in winter, but if a farmer put an excessive amount of nitrate-containing fertiliser on a neighbouring field there would be a sudden increase in any river into which the field drained.

(iii) Decide on the total number of samples you are taking, remembering that each location should be sampled in duplicate. Although it is good practice to start by taking as many samples as you feel necessary, for complete monitoring you also have to take into consideration the time required for the analyses. It is very common to underestimate severely the time involved in the laboratory analyses.

A further consideration if there is to be any statistical treatment of results is that there are sufficient samples for the treatment to be significant.

(iv) Decide on the location of the sampling. If you are to take samples regularly from one location, the first consideration must be ease of access. Remember that the weather may not always be perfect. A river should ideally be sampled in the most representative area (in the main flow of the river and underneath the surface at similar depths for each sample), but of course this is not always practically possible. If you are monitoring the effect of a discharge into a river, samples should be taken far enough downstream for the discharge to be completely mixed (Fig. 3.2a). Samples taken further upstream would be unrepresentative as the analysis would depend on how much the discharge had mixed with the river.

(v) Decide on the sample volume and the sample containers. The containers are usually made of glass or polyethylene. These materials (and the container top) are not as inert as you may think. Polyethylene containers may leach organic compounds into the sample and glass bottles inorganic species (sodium, silica and other components of the glass). How much you fill the container is also important. If you are analysing volatile material or dissolved gases the container must always be full. For other components it is beneficial not to fill the container completely as the contents can then be more easily mixed before analysis. Try attempting to mix the contents of a completely full container!

(vi) Decide on the method of storage of samples. Standard methods are available for most components to minimise analyte loss. The method varies according to the physical and chemical properties of the species. For example:

Nitrate—store at 4 °C to minimise biological degradation.

Pesticides—store in the dark to avoid photochemical decomposition.

Metal ions—acidify the sample to prevent adsorption of metal ions on the sides of the container.

Phenols—add sodium hydroxide to lower the volatility.

After all these considerations, you can start sampling!

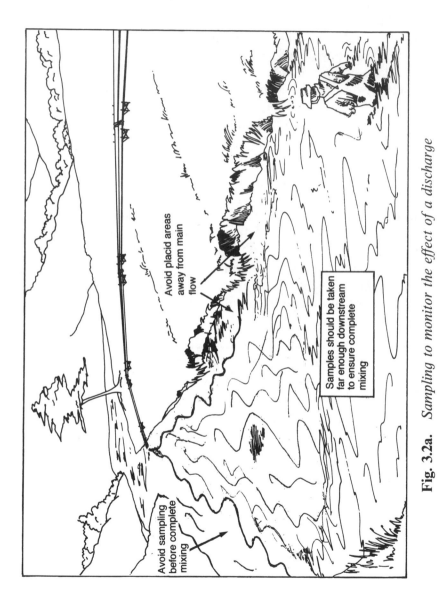

Avoid placid areas away from main flow

Samples should be taken far enough downstream to ensure complete mixing

Avoid sampling before complete mixing

Fig. 3.2a. *Sampling to monitor the effect of a discharge*

SAQ 3.2a

You are about to take samples for the following analyses:

Ammonia

Chloroform

Total organic content

List decisions which have to be made in developing a river sampling protocol.

What are the relevant chemical and physical properties which would help you decide storage conditions? Suggest storage bottles and precautions to minimise analyte loss.

3.3. MEASUREMENT OF WATER QUALITY

This section contains techniques which are usually intended to provide a measurement relating to the overall effect of groups of compounds or ions rather than measuring concentrations of individual components. Measurements that we shall be looking at include water hardness, suspended solids, oxygen demand (measuring oxidisable material), total organic carbon and electrical conductivity.

3.3.1. Suspended Solids

We can all visualise streams so full of suspended material that the water is opaque, and where no visible life could possibly exist. This represents an extreme case of high solids loading. Any natural water will contain some suspended solids, but often the material is of such a small particle size that it cannot be easily seen. It is only when you look at two samples of water, one of which you consider 'clean' and the other which has been filtered to the sub-micrometre level that you can see the difference. The filtered water glistens and the 'clean' water suddenly looks distinctly dirty. Even if the particles are chemically inert their physical properties could cause problems.

∏   What physical problems do you think may be caused by suspended solids?

1. They cut down light transmission through the water and so lower the rate of photosynthesis in plants.

2. In less turbulent parts of the river some of the solids may sediment out, smothering life on the river bed.

You may have guessed that analysis of suspended solids is by filtration and weighing but you may not realise the laboratory skill which is required until you discover that a typical suspended solid loading for a clean-looking stream would be only a few $mg\,l^{-1}$. Even sewage discharges in the UK have to conform to conditions with a maximum of $30\,mg\,l^{-1}$ (after tenfold dilution of the discharge).

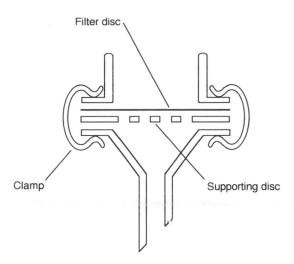

Fig. 3.3a. *A Hartley funnel*

Typically a glass-fibre filter disc with a 1.6 μm pore size would be used with a Hartley filter funnel (Fig. 3.3a). The paper is clamped inside the funnel and this prevents any part of the sample from escaping around the side of the filter.

3.3.2. Dissolved Oxygen and Oxygen Demand

All animal life in a river is dependent on the presence of dissolved oxygen. More subtle requirements for a healthy river also necessitate the presence of oxygen.

We have already seen how the presence of organic matter can remove oxygen from water by oxidation. Although the process can be written down as a simple chemical reaction, it is in fact a microbiological process, known as aerobic decay. This converts the major elements present in plant matter (C, H, N, S) into CO_2, H_2O, NO_3^- and SO_4^{2-}.

It will perhaps come as a surprise that even if no oxygen is present in the water, organic material will still be broken down. Instead of the material being oxidised it is reduced. The process is once again microbiological and is known as anaerobic decay. In this case the final products are CH_4, NH_3 and H_2S.

Consideration of the products of anaerobic decay show that this condition should be avoided at all costs in environmental waters (consider their flammability, toxicity and smell).

Oxygen is found in rivers as a result of photosynthesis by plants and by dissolution of oxygen from the atmosphere into the water. The solubility of oxygen in water is low. Water saturated with oxygen at 25 °C contains 8.24 mg l^{-1}. It would therefore be easy to deplete the oxygen content if any material were present which would react rapidly with the oxygen. The material could be organic, as already discussed, but could also be inorganic. Iron in the form of Fe^{2+} can deplete oxygen by oxidation to Fe^{3+}.

Natural replenishment by oxygen from the atmosphere can be very slow.

∏   Which of the following rivers do you think would take up oxygen most quickly? Which would be likely to have the highest oxygen demand?

1. A fast-flowing mountain stream.

2. A slowly flowing river in a heavily industrialised area.

3. A slowly flowing river in unspoilt countryside.

The turbulence caused by the fast flow cascading over rocks in the mountains would ensure that oxygen was taken up rapidly and the water saturated with oxygen. It would be unlikely that the river would contain large quantities of organic matter either from vegetation or from industrial effluent. The oxygen demand would be low.

The slowly flowing rivers would take up oxygen more slowly as there would be much less turbulence. The heavy industry would be very likely to discharge oxygen-consuming effluent which would increase the oxygen demand of the receiving water.

The river in the countryside would be less likely to contain oxygen-consuming effluent but may still possess a significant oxygen demand from decaying vegetation and also from any material carried downstream into the area.

You should be able to recognise two distinct analyses which could be useful if monitoring environment waters for oxygen.

1. A direct measurement of the oxygen concentration in the sample. This would give an indication of the health of the river at a particular location and at the time of sampling. It would be of less use for assessing the overall health of a river as the oxygen level can vary dramatically with location and with time.

2. A measurement of the amount of material which, given time, could deplete the oxygen level in the river. This is known as the oxygen demand. It gives an indication of the possibility of oxygen deple tion which will occur if the oxygen is not replenished.

   Such a measurement would be much more suitable for determining the overall health of the river since the oxygen demand of a water supply is unlikely to change suddenly.

The analytical techniques used for dissolved oxygen measurement can also be used to measure oxygen demand, so these will be discussed first.

Dissolved Oxygen

The determination of oxygen can be either by titration (Winkler method) or by the use of an electrode sensitive to dissolved oxygen. The results are either expressed as a simple concentration ($mg\, l^{-1}$) or as a percentage of full saturation. The concentration of oxygen in saturated water is dependent on temperature, pressure and salinity of the water and would need to be established from published tables or determined experimentally. The first problem to overcome is transport of the sample to the laboratory. Without modification of the sample this would cause sufficient agitation to the water to saturate the sample with oxygen from the air, regardless of its original content.

In the Winkler method the oxygen is 'fixed' immediately after sampling by reaction with Mn^{2+}, added as manganese(II) sulphate, together with an alkaline iodide/azide mixture:

$$Mn^{2+} + 2OH^- + \tfrac{1}{2}O_2 \longrightarrow MnO_2(s) + H_2O$$

The iodide is necessary for the analytical procedure in the laboratory and the azide is present to prevent interference from any nitrite ions which can oxidise the iodide.

The sample completely fills the bottle to ensure no further oxygen is introduced. After transport to the laboratory the sample is acidified with sulphuric or phosphoric acid. This results in the reaction

$$MnO_2 + 2I^- + 4H^+ \longrightarrow Mn^{2+} + I_2 + 2H_2O$$

The released iodine can then be titrated with sodium thiosulphate using a starch indicator:

$$I_2 + 2S_2O_3^{2-} \longrightarrow S_4O_6^{2-} + 2I^-$$

∏   What is the equivalence between the original oxygen and the thiosulphate?

The overall reaction is

$$2S_2O_3^{2-} + 2H^+ + \tfrac{1}{2}O_2 \longrightarrow S_4O_6^{2-} + 3H_2O$$

i.e. 4 mol of thiosulphate in the final titration is equivalent to 1 mol of oxygen in the sample.

The electrode method is used for field measurement of dissolved oxygen and can also be used in the laboratory for determination of biochemical oxygen demand. Several types are available, including the Mackereth cell shown in Fig. 3.3b.

The current generated by the cell is proportional to the rate of diffusion of oxygen through the membrane, which is in turn proportional to the concentration of the oxygen in the sample. The reactions involved are as follows:

At the cathode:

$$\tfrac{1}{2}O_2 + H_2O + 2e \longrightarrow 2OH^-$$

At the anode:

$$Pb + 2OH^- \longrightarrow PbO + H_2O + 2e$$

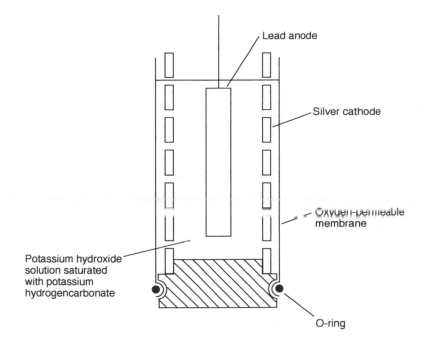

Fig. 3.3b. *A Mackereth cell*

Instruments usually read oxygen directly with a scale from 0 to 100% saturation by setting 100% with fully aerated water and 0% with water with no oxygen content (sodium sulphite is added to the water). This calibration must be made each time the electrode is used.

Oxygen Demand

This can be measured by a number of methods. We shall compare them after each method has been described.

Biochemical Oxygen Demand (BOD)

This method attempts to replicate the oxidation conditions found in the environment. The dissolved oxygen level of a fully aerated water sample is measured by either of the methods already described. The measurement is repeated on a sample after it has been left for 5 days in the dark under standard conditions which are designed to be ideal to promote microbiological activity. The sample is kept in an incubator at 20 °C after adjustment of the pH to between 6.5 and 8.5.

If the sample is expected to have a high oxygen demand, a dilution should be made. The diluent may include salts containing magnesium, calcium, iron(III) and phosphate as nutrients. A seed sample of sewage may be added if for any reason the sample is thought to be sterile.

If there is no dilution of the sample,

BOD =
(initial oxygen concentration − final oxygen concentration) mg l^{-1}

Typical BOD values for unpolluted water are of the order of a few mg l^{-1}. Many seemingly innocuous effluents have a very high oxygen demand, as shown by Fig. 3.3c.

If you remember that the saturated oxygen level in water is of the order of 8 mg l^{-1} then you will be able to see how the introduction of a small quantity of high-strength effluent can deplete the oxygen in many times its own volume of water.

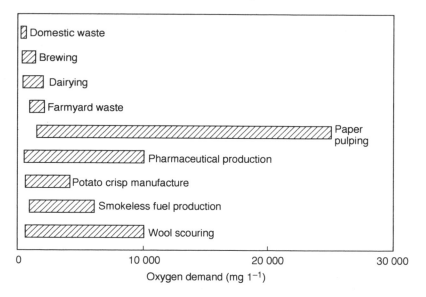

Fig. 3.3c.   *Typical effluents with high oxygen demands*

Chemical Oxygen Demand (COD)

This term covers a family of techniques which involve reacting the sample with excess of oxidising agent. After a fixed period the concentration of unreacted oxidising agent is determined by titration. The quantity of oxidising agent used can be calculated and the oxygen equivalent determined.

Chemical oxygen demand methods include the following:

Two hour dichromate value. The sample is refluxed with excess of potassium dichromate in concentrated sulphuric acid for 2 h:

$$Cr_2O_7^{2-} + 14H^+ + 6e \longrightarrow 2Cr^{3+} + 7H_2O$$

Silver sulphate may be included to catalyse the oxidation process of alcohols and low molecular mass acids.

Chloride ions give a positive interference by the reaction

$$Cr_2O_7^{2-} + 6Cl^- + 14H^+ \longrightarrow 2Cr^{3+} + 3Cl_2 + 7H_2O$$

The interference is reduced by addition of mercury(II) sulphate, a chloro complex being formed.

The excess of dichromate is titrated with ammonium iron(II) sulphate:

$$6Fe^{2+} + Cr_2O_7^{2-} + 14H^+ \rightarrow 6Fe^{3+} + 2Cr^{3+} + 7H_2O$$

∏   Given that oxidation by oxygen can be represented by

$$O_2 + 4H^+ + 4e \longrightarrow 2H_2O$$

What is the equivalence between dichromate and oxygen?

1 mol of $Cr_2O_7^{2-}$ consumes 6 mol of electrons to produce 2 mol of Cr^{3+}. Since each mole of O_2 can consume 4 mol of electrons to make H_2O, then 1 mol of $Cr_2O_7^{2-}$ is equivalent to 1.5 mol of O_2.

Permanganate tests. Excess of potassium permanganate is added under specified conditions which can range from 3 min on a steam-bath to 4 h at room temperature. The unreacted permanganate can be determined by numerous methods, including the liberation of iodine, followed by titration of the iodine with thiosulphate;

$$2MnO_4^- + 16H^+ + 10I^- \longrightarrow 2Mn^{2+} + 8H_2O + 5I_2$$
$$I_2 + 2S_2O_3^{2-} \longrightarrow S_4O_6^{2-} + 2I^-$$

The confusing number of variations of this method leads to limitation in its use since it is very difficult to obtain comparative inter-laboratory data. The 3 min variant of the test does, however, provide a rapid method of testing a water for oxidising ability.

Comparison of Tests

BOD   Closely related to natural processes.

                                5 day analysis time.

                                Difficult to reproduce, both within and between laboratories.

Chemical tests (COD)   Less relationship to natural processes.

                                Rapid analysis.

                                Good reproducibility.

                                Can analyse heavily polluted water.

All of the chemical tests will be affected by the presence of inorganic reducing or oxidising agents, the former giving positive results and the latter possibly negative results.

∏   On the basis of the comparisons suggest appropriate applications of the techniques.

BOD   Long-term monitoring of natural water.

COD   Rapid analysis of heavily polluted samples, e.g. industrial effluents.

Relationship of Oxygen Demand to Specific Concentrations

If a single organic compound was present in the water and the oxidation reactions proceeded to completion, the above methods would give an accurate measurement of its concentration. The determination of known amounts of a single compound can be used in the laboratory to test experimental procedures. Potassium hydrogenphthalate is often used. This is oxidised according to the following equation:

$$C_8H_5O_4K + 7\tfrac{1}{2}O_2 \longrightarrow 8CO_2 + 2H_2O + K^+ + OH^-$$

∏ What is the COD of a solution containing 0.340 g l^{-1} potassium hydrogenphthalate?

Relative molecular mass of potassium hydrogenphthalate = 204.

In 1 litre of solution there are 0.340/204 mol.

1 mol of potassium hydrogenphthalate = 7.5 mol of oxygen.

0.340/204 mol of potassium hydrogenphthalate
$$= 7.5 \times 0.340/204 \text{ mol of oxygen}$$
$$= 7.5 \times 0.340/204 \times 32 \times 1000 \text{ mg of oxygen}$$
$$= 400 \text{ mg.}$$

Hence COD = 400 mg l^{-1}.

3.3.3. Total Organic Carbon (TOC)

None of the methods described so far for oxygen demand gives a precise estimation of the total organic loading of the water. A number of instruments are now available which can achieve this. All involve the oxidation of the organic matter to carbon dioxide, after prior acidification to remove interference from carbonates. Methods used include the following:

(a) Injection of a small quantity of water into a gas stream passing through a heated tube to perform the oxidation. Measurement is possible down to the mg l^{-1} level.

(*b*) Wet oxidation by using potassium peroxodisulphate at room or elevated temperatures. This method is about 100 times more sensitive than heated-tube oxidation.

The carbon dioxide can then be measured either by absorption in solution and measurement of the conductivity of the solution, reduction to methane and analysis of this gas by flame ionisation detection (Section 4.2) or direct measurement by infrared spectrometry (Section 6.3).

Current trends are to replace BOD and other oxygen demand measurements with TOC. To understand this you should note the following advantages:

 (i) it is a rapid technique;

(ii) it would be expected to give highly reproducible results;

(iii) it can be easily automated, either for laboratory analysis or for on-line monitoring of effluents.

3.3.4.   pH, Acidity and Alkalinity

pH is related to the number of hydrogen ions in solution by the relationship

$$pH = -\log_{10}a(H^+)$$

where $a(H^+)$ is the hydrogen activity. At the low concentrations of hydrogen ions which are typically found in the environment the hydrogen ion activity is approximately equivalent to the hydrogen ion concentration.

Typical pH values found with environmental water samples are shown in Fig. 3.3d.

Did you realise that unpolluted rainwater is slightly acidic? This is due to the presence of dissolved carbon dioxide:

$$H_2O + CO_2(gas) \rightleftharpoons H_2O{\cdot}CO_2(solution) \rightleftharpoons$$

$$H^+ + HCO_3^- \rightleftharpoons 2H^+ + CO_3^{2-}$$

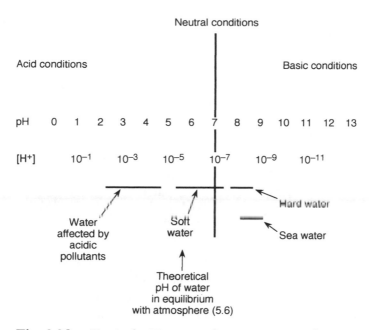

Fig. 3.3d. *Typical pH ranges for environmental waters*

Hard water is slightly alkaline. The hardness is due to the presence of polyvalent metal ions, e.g. calcium and magnesium, arising from dissolution of minerals. For instance, the dissolution of limestone involves the following equilibria:

$$CaCO_3 \rightleftharpoons Ca^{2+} + CO_3^{2-}$$
$$CO_3^{2-} + H_2O \rightleftharpoons HCO_3^- + OH^-$$

The biological effect of changes in pH can most easily be seen by the sensitivity of freshwater species to acid conditions. Populations of salmon start to decrease below pH 6.5, perch below pH 6.0 and eels below pH 5.5, with little life possible below pH 5.0. The eradication of life can result from a change of little more than 1 pH unit.

There are also chemical effects. In Section 2.4 we discussed how a decrease in pH increases the solubility of metals. The use of lead piping for domestic water supplies becomes of greater concern as the water becomes more acidic. The weathering of minerals such as limestone or dolomite by water becomes more rapid with a decrease in pH.

A typical procedure for measurement of pH involves calibration with two buffer solutions spanning the expected pH of the sample, then measurement of the sample.

The procedures for 'acidity' and 'alkalinity' measure, by titration, the quantity of base or acid needed to change the pH of a sample to a fixed value (usually 4.5 or 8.3), corresponding to methyl orange and phenolphthalein end-points, respectively. From a chemical point of view this gives a measurement of the buffer capacity (resistance to change in pH) of the water. A high buffer capacity is a useful feature if an acidic or basic pollutant is being added to the water.

3.3.5. Water Hardness

The term 'water hardness' will be very familiar to you if you live in an area where there are high concentrations of calcium and magnesium in your water supply. The effects you may have noticed include:

1. Deposition of a white solid whenever the water is heated. This is commonly seen as 'furring' of kettles. This may also lead to blockage of hot-water pipes.

2. The formation of scum whenever soap or washing powder is added to water.

A small degree of hardness does, however, have some beneficial effects. The alkalinity lowers the solubility of toxic metals. The buffering action of the salts producing the hardness lessens the effect of acidic pollutants. In addition, there is evidence that hard water is beneficial to health, particularly in the reduction of heart disease, and it certainly is more pleasant to drink.

The effects are generally produced by the presence of polyvalent metal ions in the water from the weathering of minerals. This usually is almost entirely due to calcium and magnesium ions, but others such as aluminium, iron and zinc ions may make a small contribution. The minerals are often based on carbonates (limestone, $CaCO_3$; dolomite, $CaCO_3 \cdot MgCO_3$) or sulphate (gypsum, $CaSO_4$). It is only the hardness derived from carbonates which gives rise to solid deposition ('carbonate' hardness or 'temporary' hardness). Hardness which does not produce this effect is 'non-carbonate' or 'permanent' hardness.

Analysis is normally performed by complexometric titration using the disodium salt of ethylenediaminetetraacetic acid (EDTA).

$$HO_2C-CH_2 \diagdown \atop HO_2C-CH_2 \diagup N-CH_2-CH_2-N \diagup CH_2-CO_2H \atop \diagdown CH_2-CO_2H$$

EDTA

This forms a 1:1 complex with divalent metal ions:

$$M^{2+} + EDTA^{2-} \rightleftharpoons M(EDTA)$$

To determine both calcium and magnesium by titration the pH has to be buffered at 10 [a higher pH would precipitate magnesium ion as $Mg(OH)_2$]. The end-point is detected using an indicator such as Erichrome Black T.

The titration estimates the *total* divalent metal as a molar concentration. Many non-chemists are unfamiliar with molar concentrations and so the quantity is often re-expressed in more familiar terms. However, it would be impossible to convert the value into the more familiar mass concentrations ($mg\,l^{-1}$) without knowing the precise individual concentrations of calcium, magnesium and other ions. Even then you would not be able to quote a single figure for the total hardness, just a table of individual concentrations.

In order to overcome this, the total hardness is expressed in $mg\,l^{-1}$ units as if it were all calcium carbonate, even if it is due to calcium sulphate, magnesium carbonate or any other polyvalent metal salt.

∏   Which of the following solutions give $50\,mg\,l^{-1}$ total hardness?:

(a) $50\,mg\,l^{-1}$ $MgCO_3$;

(b) $21.1\,mg\,l^{-1}$ $MgCO_3$ + $25\,mg\,l^{-1}$ $CaCO_3$;

(c) $50\,mg\,l^{-1}$ $CaSO_4$;

(d) $55\,mg\,l^{-1}$ $CaCl_2$.

The concentrations of the above, expressed as molarities, are (a) 0.60, (b) 0.5 (0.25 + 0.25), (c) 0.37 and (d) 0.5 mM.

The hardness can be determined by multiplying by the relative molecular mass of calcium carbonate (= 100). This gives the total hardness of the solutions as (a) 60, (b) 50, (c) 37 and (d) 50 mg$\,l^{-1}$.

What values should we expect from environment samples? Although the terms 'hard' and 'soft' sound very subjective, they have to come to be defined within very specific concentration ranges. Within the UK the following definitions are used:

mg l^{-1} CaCO$_3$	
0–50	soft
50–100	moderately soft
100–150	slightly hard
150–200	moderately hard
200–300	hard
> 300	very hard

3.3.6.   Electrical Conductivity

You may sometimes wish to know the total inorganic salt content in a sample. A simple method would be to evaporate the sample to dryness and weigh the solid. Large volumes of sample would need to be evaporated, however, making the technique less attractive than at first thought.

It would be much more convenient if an electrode could simply be placed in the sample to make the measurement. The closest method to this ideal is the use of a conductivity cell for dissolved ions, as illustrated in Fig. 3.3e.

A low-voltage alternating current is applied across the electrodes. The resistance of the liquid between the electrodes is measured, which is

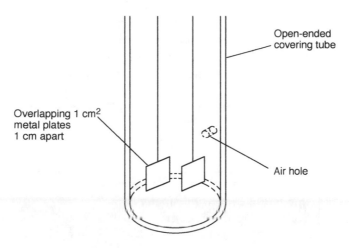

Fig. 3.3e. *A conductivity cell*

converted into conductivity according to the equation

$$K = L/AR$$

where

K = conductivity;

L = distance between electrodes (cm);

A = surface area of electrodes (cm^2);

R = resistance (siemens, S).

The units of conductivity applicable to environmental samples are μS cm^{-1}, a typical value of 200 μS cm^{-1} being found for water with a substantial ionic salt content.

The cell is calibrated using solutions of known conductivity. Conductivity is highly temperature dependent and so care has to be taken that calibration solutions and the unknown sample are at the same temperature.

The relationship between conductivity and total salt content is not simple. All ions having the same charge have approximately the same

conductivity, but unfortunately most environmental waters contain ions
with different charges in varying concentrations. If a series of waters of
roughly similar composition are known, an approximate conversion can
be made. For many waters in the UK the following equation is valid:

$$\text{total salt concentration} = A \times \text{conductivity (mg l}^{-1})$$

where A is a constant in the range 0.55–0.80.

SAQ 3.3a

The UK classification of rivers includes the following
limiting chemical parameters:

Class	Dissolved oxygen (%)	Biochemical oxygen demand (mg l^{-1})	Ammonia (mg l^{-1})	Potential Use
1A	>80	<3	<0.4 ⎫	⎧ Drinking water
1B	>60	<5	<0.9 ⎭	Amenity ⎫
2	>40	<9		Drinking water after treatment Coarse fishing
3	>10	<17		Low-grade industrial purpose
4	<10			No industrial use

Why do you consider these parameters and limiting
concentrations are used

SAQ 3.3a

3.4. TECHNIQUES FOR THE ANALYSIS OF COMMON IONS

This section discusses the application of techniques to determine ions present in the $mg\,l^{-1}$ range; (which are they? check with Section 3.1). You will begin to note that from this section onwards the techniques become almost entirely instrumental, confirming one of the major advantages of these methods, i.e. the ability to analyse low concentrations in multi-analyte samples with ease. You will find that many of the instrumental methods for ions within this concentration range need little sample preparation. Later when we discuss the analysis of ions at $\mu g\,l^{-1}$ levels much of our discussion will concern the preconcentration of samples to bring them within the working range of the instruments. The instrumental method then becomes just one part of the more complex analytical procedure.

3.4.1. Visible and Ultraviolet Spectrometry

∏   From your knowledge of the technique, describe the law on which the analytical method is based.

At sufficiently low concentrations the Beer–Lambert law is followed:

$$A = \varepsilon c l$$

where

A = absorbance of radiation at a particular wavelength;
= $\log (I_0/I)$;

I_0 = intensity of incident radiation;

I = intensity of transmitted radiation;

ε = proportionality constant (molar absorptivity; $1 \, \text{mol}^{-1} \, \text{cm}^{-1}$);

c = concentration of absorbing species ($\text{mol} \, 1^{-1}$);

l = path length of light beam (cm).

If you had difficulty remembering this law, then revise your knowledge before proceeding. The Beer–Lambert law is fundamental to many of the techniques we shall be discussing.

The instruments used to measure the absorption of light can range from sophisticated laboratory instruments which can operate over the whole visible–ultraviolet range to portable colorimeters using natural visible light, which are used as field instruments. This makes absorption spectrometry one of the most useful and versatile techniques for an environmental analyst.

You might at first hesitate to believe the last statement. After all, none of the common ions in water absorb light in the visible region of the spectrum (natural water is usually almost colourless). Also, the only ions commonly found in water which absorb in the ultraviolet range above 200 nm are nitrate and nitrite.

The main use of the technique involves the production of light-absorbing derivatives of these ions. This can be performed for almost all the common anions (except sulphate) and also ammonia.

Analysis by direct absorption:

nitrate

Analysis after formation of derivative:

chloride

fluoride

nitrate

nitrite

phosphate

As an example of the technique, the procedure for phosphate involves the addition of a mixed reagent (sulphuric acid, ammonium molybdate, ascorbic acid, antimony potassium tartrate) to a known volume of sample, diluting to volume, shaking and leaving for 10 min. A blue phosphomolybdenum complex is produced, the absorbance of which is measured. The concentration is calculated using a predetermined calibration graph derived from standard solutions treated in the same way.

Automatic procedures have been devised for most of the ions listed above.

A typical apparatus is shown in Fig. 3.4a.

Instead of mixing reagents for each analysis, streams of each reagent (segmented by air bubbles to diminish mixing effects) in narrow-bore tubes are mixed by combining the flows at a T-junction or within a mixing cell. A sample is introduced from an automatic sampler as a continuous flow into the reaction stream. The combined flow is then led into a spectrophotometer and the absorption measured. The flows of all the reagents and samples are produced from a multi-channel peristaltic pump (Fig. 3.4b).

Field Techniques

Field techniques are becoming increasingly important to give immediate measurement of ion concentrations. Unmanned field stations can be set up using the automatic procedures described above. Alternatively, portable (often hand-held) instruments may be used.

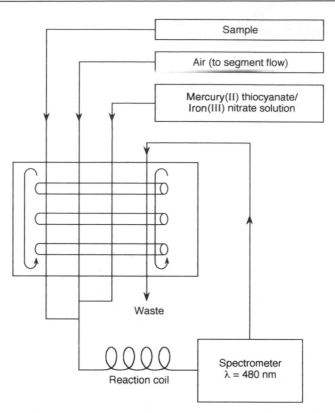

Fig. 3.4a. *Continuous-flow analysis of chloride*

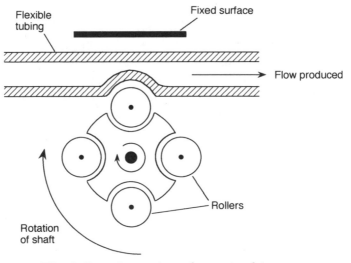

Fig. 3.4b. *Operation of a peristaltic pump*

∏ What modifications need to be made to the standard apparatus and methods already described for portable field instruments?

The procedure for the colour-forming reaction has to be made simple. No-one wishes to perform complicated analytical routines on a muddy riverbank!

Calibration of the instrument should avoid the use of standard solutions, which again are inconvenient in the field.

The optical components of the instrument should be minimised or, at least, be made robust.

Each manufacturer has a different approach to these modifications. Colour-forming reagents may be premeasured in the form of tablets or in solution. As a further simplification, one manufacturer seals the reagents under vacuum in an ampoule. Breakage of the top under water automatically draws the correct sample volume into the ampoule.

Coloured glass or moulded plastic standards are often used rather than solutions. These can be in the form of discs. One manufacturer's design contains glasses of different optical density (Fig. 3.4c). The disc is rotated

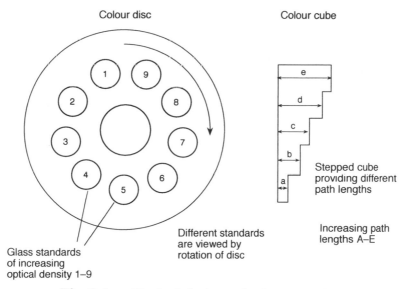

Colour disc

Colour cube

Glass standards of increasing optical density 1–9

Different standards are viewed by rotation of disc

Stepped cube providing different path lengths

Increasing path lengths A–E

Fig. 3.4c. *Typical designs of colour standards*

through the light beam until the colour of the standard glass matches that of the unknown. Alternatively, a moulded plastic cube may be used which has a stepped side to provide a number of possible path lengths (and hence absorbances).

The simplest procedure for quantification is by visual comparison of the colour of the standards and the unknown using available sunlight. A hand-held instrument has been described in another volume in this series (*Visible and Ultraviolet Spectroscopy*, Fig. 1.2d). Alternatively portable spectrometers are available, often housed in briefcases along with titration apparatus and pH and conductivity electrodes and are known as 'water quality' test kits.

3.4.2.   Emission Spectrometry (Flame Photometry)

Emission spectrometry relies on the principle that for some metals at low concentrations, the intensity of light emitted from an electronically excited atom (usually produced by introduction of the sample into a flame) is proportional to the concentration of the excited species. Simple and inexpensive instrumentation is available, often described by the term 'flame photometry.'

∏   List some of the reasons why the technique seems almost ideally suited to analysis of environmental samples.

1. Although the use of flame photometry is limited to a few alkali and alkaline earth metal ions, this includes sodium, potassium and calcium, three of the four major cations in water (What is the fourth? Check in Section 3.1 if you are unsure.)

2. The linear concentration range (for sodium and potassium $1–10$ mg l^{-1} and for calcium $10–50$ mg l^{-1}) is within the range expected for environmental water samples. Little sample preparation is needed.

3. The instrument is simple to use and the only laboratory requirements are a gas supply (natural gas is adequate) and a source of vacuum. It can be easily installed in temporary laboratories for analysis close to the sampling site.

It is a pity that flame photometers cannot be used to analyse the fourth common ion, magnesium, as all the routine analytical requirements for metal ions could then be performed by this simple method. Analysis for magnesium is usually carried out by atomic absorption spectrometry.

The major disadvantage of flame photometry is the variation of response of the instrument with time (i.e. drift). Great care has to be taken that calibration of the instrument and analytical measurements are performed quickly after each other. It is also good practice to repeat the calibration after the analysis to check that no variation has occurred.

3.4.3. Ion Chromatography

The methods that we have looked at so far have been for the analysis of individual ions, but sometimes a complete analysis of all the ions in the sample is needed. Chromatographic separation of the ions is an obvious approach. Liquid chromatography would seem particularly useful since the species to be analysed are already in solution. From your reading elsewhere you will be familiar with the principles of high-performance liquid chromatography (HPLC) and how its application over the last two decades has expanded to include virtually all soluble ions and compounds. The major application in environmental analysis has been for inorganic anions. Several variations of the liquid chromatographic technique have been developed using either conventional high-performance liquid chromatographs or specialised 'ion' chromatographs. Since you will most likely not be familiar with the specialised technique this will be described first, and then later compared with conventional HPLC.

The components of a typical 'ion' chromatograph are shown in Fig. 3.4d.

The separation of the anions is achieved using a polystyrene-based ion-exchange column with an eluent typically containing a sodium carbonate/hydrogencarbonate buffer. Detection of the analyte ions is by monitoring the increase in conductivity of the eluent produced by the ions as they pass through the detector. In order to maximise the detection sensitivity, prior to passing to the detector all buffer ions have to be removed from the eluent (these would contribute to the back-

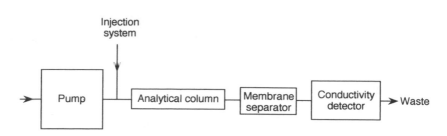

Fig. 3.4d. *Components of a typical ion chromatograph*

ground conductivity). The sodium ions in solution are replaced with hydrogen ions. The carbonate and hydrogencarbonate react with the hydrogen ions to form carbon dioxide, which has little conductivity in solution.

$$HCO_3^- + H^+ \rightleftharpoons H_2O + CO_2$$
$$CO_3^{2-} + 2H^+ \rightleftharpoons H_2O + CO_2$$

The most widespread method of achieving this (the method has been improved and updated several times) is to use a 'micromembrane suppressor' as illustrated in Fig. 3.4e. The eluent flows between two semi-permeable membranes which separate it from a counter flow of sulphuric acid.

The membrane only allows the passage of cations. Hydrogen ions migrate into the eluent as the sodium ions migrate into the sulphuric acid. The migration of each ion is determined by the relative concentrations of the ion in the two liquids, the ion moving into the solution of lower concentration.

A recently introduced development is to use a suppression system which does not require a separate flow of sulphuric acid. Eluent is recycled on the outside of the membrane, the ion migration being provided by an electrical potential being set up across the cell.

For ions of interest in environmental water found at $mg\,l^{-1}$ concentrations, the sample would need to be diluted before injection. This, along with filtration, is often the only sample preparation necessary. Common ions in water can be determined within a few minutes (Fig. 3.4f).

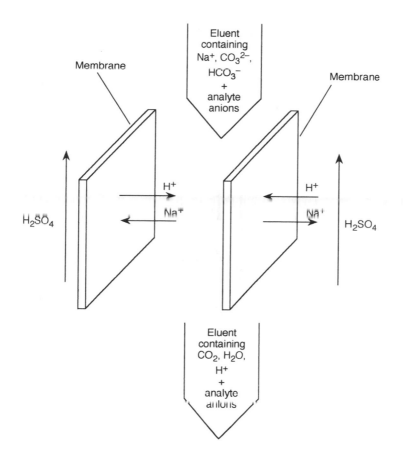

Fig. 3.4e. *Micromembrane suppression*

Π   What disadvantage can you see in using ion chromatography
     for a single ion such as chloride or nitrate?

The time taken for the analysis will be determined by the elution time
of the slowest component (often sulphate or phosphate), rather than the
component of interest. A second analysis cannot proceed until all the
ions have eluted. This makes the technique slow in comparison with the
methods dedicated to single ions, e.g. continuous-flow analysis.

Methods developed for conventional HPLC can use either a reversed-
phase column and ion-pair techniques, or an ion-exchange column.
Ultraviolet absorbance and conductivity detectors are used.

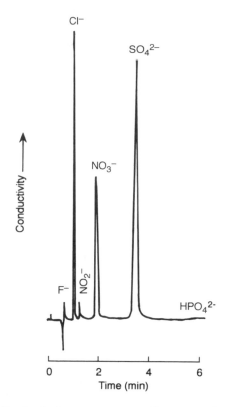

Fig. 3.4f. *Typical chromatogram of a natural water sample*

When a conductivity detector is used the system becomes similar to the specialised chromatograph already described, but without eluent ion suppression. The sensitivity is lower in comparison with the specialised chromatograph but is still sufficiently high to analyse common anions at mg l^{-1} concentrations.

⊓   What common anions can be detected using UV absorbance?

If you re-read Section 3.4.1 you will find that nitrate and nitrite ions absorb in the UV region of the spectrum. HPLC with UV detection is a useful method for these two ions, when analyses of the other ions are not required.

Although the most common use of chromatography is for anions, similar methods are now available for common cations (Na^+, K^+, NH_4^+,

Ca^{2+}, Mg^{2+}) using both the ion suppression system and also conventional chromatographs with conductivity detection. Convenient methodologies have become available only over the past few years, so applications are lagging behind those for anions.

3.4.4. Examples of the Use of Other Techniques

We have now covered the most widely used methods for analysis of the common ions. There are, however, a few frequently used methods which have not been covered. We shall look at the analysis of three species— ammonia, fluoride and sulphate.

Ammonia

Ammonia is the only alkaline gas commonly found in environmental water. If isolated from solution, the ammonia could be determined by a simple acid–base titration. This may be done by addition of magnesium oxide to ensure that the sample is slightly alkaline. The ammonia is then present predominantly in the form of NH_3, rather than the less volatile NH_4^+. Ammonia is then distilled off into pure water and titrated with standard acid. The apparatus used is shown in Fig. 3.4g.

For rapid screening of samples it is possible to use an ion-selective electrode (i.e. an electrode whose potential, measured with respect to a reference, is proportional to the logarithm of the activity of one particular ion). Although this may appear to be a new technique for you to learn, you are already familiar with one particular ion-selective electrode. A combination pH electrode is simply an ion-selective electrode responsive to hydrogen ions and a reference electrode housed in a single body. Ion-selective electrodes are available for most common ions and gases which dissolve as ionic species, but they have limitations.

Many are prone to interference from other species and thus have poor precision. Even the pH electrode has taken many years of development to produce a reliable response. All respond to ionic activity rather than concentration, so it is essential to add a large excess of an ionic salt to both the standard solutions and the unknown in order that the ionic strength of each solution is identical.

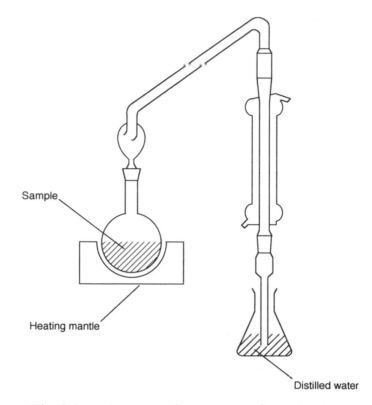

Fig. 3.4g.   *Apparatus for ammonia determination*

Ammonia electrodes are of the gas-sensing type. The ammonia diffuses through a permeable membrane and causes a pH change in a small volume of internal solution which is sensed by a 'glass' electrode. Prior to measurement, concentrated sodium hydroxide solution is added to the samples and standards. This serves to increase the pH to above 11 to ensure that the ammonia is in the unprotonated form, and also to provide a constant ionic strength. The ammonia electrodes responds only to gaseous alkaline gases. For most environmental applications (except in the analysis of heavily polluted water) there will then be little possibility of interference.

You may wish to compare and contrast these methods with the spectrometric method discussed in Section 3.4.1.

Fluoride

A second electrode which has found widespread use for water analysis is for fluoride ions. This is a solid-state electrode where the electrical potential is generated by migration of the ion through a doped lanthanum fluoride crystal. This again gives extremely high specificity to the analyte ion, the only pretreatment necessary being the addition of buffer solution to maintain constant pH and ionic strength.

Alternative techniques for fluoride determination are spectrometry (Section 3.4.1) and ion chromatography (Section 3.4.3).

Sulphate

There is no direct colorimetric method for sulphate and the ion-selective electrode for this ion is not very reliable; in fact, the only direct instrumental method is by using ion chromatography. Virtually every other method is based on precipitation of an insoluble sulphate. Barium or 2-aminoperimidinium (Fig. 3.4h) salts are used for the precipitation.

The precipitate may be weighed for a direct determination of the sulphate. This represents one of the few remaining applications of gravimetric analysis.

Other methods using insoluble salt precipitation are indirect, estimating the excess of the cation after precipitation of the sulphate. Excess barium may be determined by titration (which titration have you already come across which will analyse a divalent metal ion?) or by atomic absorption spectrometry (Section 4.3). Excess 2-aminoperimidinium ions may be estimated by visible spectrometry.

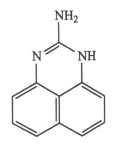

Fig. 3.4h. *2-Aminoperimidine*

SAQ 3.4a   Many of the species we have discussed can be analysed
                by more than one method. Tabulate the common
                procedures available for each of the major ions found
                in water.

SAQ 3.4b   Consider the techniques you have listed in SAQ 3.4a.
                What criteria would influence your choice of method?

Summary

The composition of water changes continuously as it travels in the environment. Sampling at a large number of locations is necessary to monitor these changes. Careful choice of locations, sampling time and sample storage procedures are necessary for reliable monitoring.

The quality of water can be assessed using measurements relating to the overall effects of groups of compounds or ions (water quality parameters) and by analysis of the major individual components. Methods for both types of determination have been discussed and include volumetric and instrumental methods.

Objectives

You should now be able to:

- list the major constituents of environmental waters and their environmental concentrations;

- appreciate the importance of correct methods of sampling and sample storage;

- describe methods for the measurement of water quality;

- determine which analytical techniques are suitable for the analysis of the major constituents of water.

SAQs AND RESPONSES FOR PART THREE

SAQ 3.1a Using the list you produced of likely chemical species in a river, decide which would be likely to increase or decrease downstream from a sampling point close to the source.

Response

Inorganic ions would be expected to increase in concentration further downstream owing to evaporation of water and from continual weathering of rocks.

The total loading of suspended solids can increase downstream as more solid material accumulates in the stream. Often, however, you will find large amounts of deposition in slow moving areas.

The build-up of organic compounds will depend on the rate at which they are being introduced into the river compared with the rate of their oxidation. The rate of oxidation will in turn depend on the rate of re-oxygenation of the water. If the re-oxygenation is rapid then the organic content will decrease; if not, the oxygen concentration will decrease and the organic content will build up.

Re-oxygenation (and hence the oxygen concentration) will be high when the river is fast flowing near its source. Further downstream, where the river is more placid, oxygen uptake will be slower. Some or, in the worst cases, all of the oxygen may be consumed by reaction with organic material.

SAQ 3.2a

> You are about to take samples for the following analyses:
>
> Ammonia
>
> Chloroform
>
> Total organic content
>
> List decisions which have to be made in developing a river sampling protocol.
>
> What are the relevant chemical and physical properties which would help you decide storage conditions? Suggest storage bottles and precautions to minimise analyte loss.

Response

Decisions have to be made concerning:

1. The analyses required.

2. The timing of the sampling (i.e. the sample programme).

3. The number of samples to be taken.

4. The location of the sampling.

5. Sample volumes and containers.

6. The method of sample storage.

Ammonia is an alkaline gas which would very easily escape from the sample. The solution could be acidified, converting the ammonia into the ammonium ion. This is less readily lost.

$$NH_3 + H^+ \rightleftharpoons NH_4^+$$

Glass or polyethylene storage containers can be used.

Chloroform is similarly volatile but it is not easy to think of a simple method of fixing it in solution. The only easy method to minimise volatilisation losses is to fill the sample container completely, store it at sub-ambient temperature and keep the storage time as short as possible. A glass container should be used. Polyethylene containers may contaminate the samples with compounds which interfere in the subsequent analysis.

Organic compounds are readily oxidised by micro-organisms using oxygen from the air. To minimise the biological activity the container should be completely filled and stored at $4\,°C$. A glass container should again be used.

SAQ 3.3a

The UK classification of rivers includes the following limiting chemical parameters:

Class	Dissolved oxygen (%)	Biochemical oxygen demand (mg l^{-1})	Ammonia (mg l^{-1})	Potential Use
1A	>80	<3	<0.4 ⎫	⎧ Drinking water
1B	>60	<5	<0.9 ⎭	⎩ Amenity
2	>40	<9		Drinking water after treatment Coarse fishing
3	>10	<17		Low-grade industrial purpose
4	<10			No industrial use

Why do you consider these parameters and limiting concentrations are used

Response

Dissolved oxygen is necessary to support animal life within the river. For an unpolluted river this should be close to saturation. Any organic pollution would tend to diminish this value.

BOD measures the oxygen-depleting potential within the river. A small quantity of material (a few mg l^{-1}) will generally be present in any river from decaying vegetation. However, once the BOD value rises above this level, there is the potential of substantially depleting the oxygen content of water. (Remember that saturated water has a concentration of 8.54 mg l^{-1} of oxygen at 25 °C.)

Ammonia is a natural constituent of water formed by the decomposition of organic material. It never reaches high concentrations under normal conditions as it is quickly oxidised to nitrate. An increase in concentration would be a further indication of the poor oxygenation of the water. High concentrations of ammonia are also toxic towards fish and would present a pollution problem in their own right.

SAQ 3.4a

> Many of the species we have discussed can be analysed by more than one method. Tabulate the common procedures available for each of the major ions found in water.

Response

Cations

	Absorption spectrometry	Ion chromatography	Titration	Flame photometry	Ion-selective electrodes
Na^+		$\times$		$\times$	
K^+		$\times$		$\times$	
Mg^{2+}		$\times$	$\times$		
Ca^{2+}		$\times$	$\times$	$\times$	
NH_4^+/NH_3	$\times$	$\times$	$\times$		$\times$

Ca^{2+} and Mg^{2+} have been included in the titration column. The standard method for water hardness gives a measurement of total polvalent ions (largely Ca^{2+} and Mg^{2+}). You should have noted that a second titration at pH > 10 will measure calcium and hence, by difference, magnesium can be estimated.

Atomic absorption spectrometry was also mentioned as often being used for magnesium analysis. As we will see in Part 4, the technique could also be used for the other metals.

Anions

	Absorption spectrometry	Ion chromatography	Gravimetric analysis	Ion-selective electrodes
Chloride	×	×		
Fluoride	×	×		×
Nitrate	×	×		
Nitrite	×	×		
Phosphate	×	×		
Sulphate	Indirect	×	×	

SAQ 3.4b   Consider the techniques you have listed in SAQ 3.4a. What criteria would influence your choice of method?

Response

The choice should depend on the following factors:

Precision of the technique—compare with precision required.

Analytical time of technique—compare with urgency of result.

Instrument time required for technique—have you sufficient instruments?

Analyst's time required for technique—the analyst's time can often be significantly different from the instrument time.

Time required to set up any instrumentation, or in preparation of reagents—this time becomes more significant with small numbers of samples.

Number of analytes to be determined—some methods can determine more than one analyte.

Availability of equipment.

Relative cost of instrumentation/labour.

All of the criteria except the last two will be independent of the country in which you work. It would be possible for an instrumental method to be favoured in one part of the world where labour costs are high whereas more labour-intensive methods are favoured in other parts of the world, and where, perhaps, instrumentation is less readily available.

4. Water Analysis— Trace Pollutants

OVERVIEW

This part of the book introduces you to the analysis of the constituents of water found at $\mu g\, l^{-1}$ concentrations, and in particular organic compounds and metal ions. The extraction and concentration procedures necessary for analysis at these concentrations are discussed.

4.1. INTRODUCTION

Before you started this course you may have thought that compounds with a concentration in water in the $\mu g\, l^{-1}$ range would have been of little environmental consequence. The introductory parts showed how certain ions and compounds could have effects significantly greater than what may have initially been expected from their environmental concentrations. These are neutral organic compounds and metal ions which readily bioaccumulate and thus are found in organisms at concentrations exceeding the background levels by many factors of ten.

Another major cause of concern is the presence in water of a number of non-bioaccumulative organic compounds which are suspected to be carcinogens. A typical example is chloroform, which can be produced in trace quantities during the disinfection of water by chlorination and which is thought to be harmful at $\mu g\, l^{-1}$ concentrations.

Until a few years ago, analysis at the lower end of these concentrations would have been beyond the capabilities of the available instrumentation and techniques, but developments since then have made these analysis routine.

The discussion of methods includes pretreatment. Pretreatment is required to remove potential interferences and, for many techniques, to increase the analyte concentration to within the instrument sensitivity. The care necessary throughout the analysis when determining compounds at these low concentrations is also emphasised.

4.2. ORGANIC TRACE POLLUTANTS

The range of organic compounds which may be found in environmental waters includes:

— Naturally occurring compounds from decaying organic material.

— Pollutants discharged into the environment.

— Degradation and inter-reaction products of the pollutants.

— Substances introduced during sewage treatment.

Typical analyses could include:

— Analysis of individual compounds or groups of compounds of environmental concern.

— Total analysis of all organic components above the limit of detection. This is an enormous task and at the lower end of the concentration range there will almost invariably be unidentified components.

— Qualitative identification of trade products in spillages or discharges.

∏ We found in earlier chapters that the properties of compounds causing widespread environmental problems include toxicity, slow biodegradation and ability to bioaccumulate within organisms. List the types of organic compound which may be included in the classification.

You should have included in your list:

pesticides, particularly those containing chlorine;

polychlorinated biphenyls;

dioxins.

For more localised pollution problems, we could extend our list of concern to include virtually every organic compound currently in use or production, together with their reaction and degradation products.

∏   Analysis of such complex mixtures would normally involve the chromatographic separation of the components. Which form of chromatography would you consider most appropriate?

As most organic compounds have significant volatility even at room temperature, gas chromatography would be expected to be a useful technique. The alternative of high performance liquid chromatography is used only where there are advantages over established gas chromatographic methods, although the number of applications of this technique is increasing.

A major area where non-chromatographic methods are used is in the determination of groups of compounds such as phenols, and also of classes of detergents, where the total concentration of the group of substances is required rather than the concentration of individual compounds.

∏   What technique have you met which could analyse related classes of organic compounds?

Visible/ultraviolet absorption spectrometry appears ideal. Absorptions are broad and the molar absorptivities often vary little between compounds within groups. A single absorption measurement could be used to determine the total concentration of the group.

Although there may be suitable volumetric techniques for individual groups of compounds they would not be sufficiently sensitive for concentrations in the $\mu g \, l^{-1}$ range.

We shall be concentrating mainly on gas chromatographic methods, later briefly discussing the other techniques.

4.2.1. Guidelines for Storage of Samples and Their Subsequent Analysis

∏   In the last chapter we covered general principles for sample storage. List the considerations necessary for organic trace pollutants.

The following list should not contain too many surprises.

(*a*) The volatility of organic compounds.

Even high relative molecular mass compounds (e.g. pesticides) have a significant vapour pressure at room temperature. Storage containers should be completely filled and kept at sub-ambient temperatures. A temperature of 4 °C is often specified in analytical procedures. This is the temperature of a normal domestic refrigerator.

(*b*) Microbial degradation.

Storage at 4 °C will lower microbial activity; storage below 0 °C (e.g. in a deep-freeze) will lower this still further.

(*c*) Photolytic decomposition.

Many potential analytes (e.g. organochlorine pesticides) are photosensitive in dilute aqueous solution. The samples should be stored in the dark.

(*d*) Contamination from the container.

Glass bottles should be used. Bottles made of organic polymers will leach potentially interfering monomers and additives into the sample.

(*e*) Loss of analyte on the container walls

Low-solubility organic compounds can be adsorbed on the container walls. This problem cannot be fully overcome. The best method of minimising the effect is to proceed with the analysis as quickly as possible.

Sample volumes required depend on the concentration of the analyte. Although the chromatographic techniques used involve the injection of just a few microlitres solution, and spectrophotometric analysis a few millilitres, this solution may first have been extracted from several litres of sample.

The precautions necessary to avoid either contamination or loss of material at these low concentrations, during the subsequent analysis, are often not appreciated. A few typical precautions indicate the caution necessary;

(*a*) The analysis should be performed in a laboratory as free as possible from the analyte. Remember many of these trace contaminants are solvents frequently found in analytical laboratories.

(*b*) Any stock solvents should be safeguarded, minimising exposure to the atmosphere and avoiding sample withdrawal with potentially contaminated pipettes or syringes.

(*c*) Samples and working standards should be placed well away from more concentrated solutions or stock solvents.

(*d*) As traces of pesticides are commonly found in laboratory solvents, pesticide-free grade solvents should be used for these analyses.

(*e*) Glassware should be scrupulously cleaned or new, if at all possible.

Such is the problem of contamination that the practical lower limits of detection can often be limited by the background concentrations of the analyte (or of interfering components) in the reagents or laboratory atmosphere.

4.2.2. Gas Chromatography

∏ From your knowledge of the technique, sketch the major components of a typical gas chromatograph. What is the principle by which the separation occurs?

The main components of a gas chromatograph are shown in Fig. 4.2a.

Chromatographic separation of a mixture occurs by the differential retention of the components between a stationary phase and a mobile phase. In gas–liquid chromatography, the mobile phase is a gas and the stationary phase is a liquid adsorbed on or chemically bonded to a solid.

Gas chromatography has the advantages over other chromatographic techniques of combining high separation efficiencies with the availability of highly specific and sensitive detectors. A high proportion of the separations required can be performed using just a few stationary phases. The wide range of phases which are available does, however, permit the development of columns for specific problem separations.

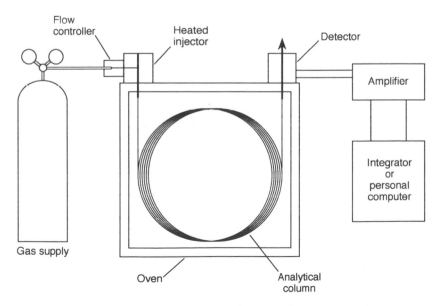

Fig. 4.2a. *Components of a gas chromatograph*

We shall discuss the columns and detectors used for water analysis first and then give examples of a complete analytical procedure, including sample pretreatment.

Detectors

The most common detectors used for environmental trace analysis are listed below together with typical applications.

Flame ionisation detector	Sensitive universal detector for organic compounds.
Electron-capture detector	Highly sensitive specific detector responding to atoms with a high electron affinity, e.g. chlorine. Typical analytes are chlorinated pesticides and chlorinated solvents.
Hall electrolytic conductivity detector	Highly sensitive specific detector for halogens, nitrogen and sulphur. Typical analytes are pesticides and trihalomethanes.
Thermionic detector	Element-specific detector for compounds containing nitrogen and phosphorus. Typical analytes are pesticides.
Flame photometric detector	Element-specific detector for compounds containing sulphur and phosphorus. Typical analytes are pesticides.
Photo-ionisation detector	Specific to compounds with aromatic rings or double bonds. Typical analytes are industrial solvents.
Mass spectrometric detector	Highly specific and sensitive detection for all organic compounds. It can also be used for peak identification.

The electron-capture detector has held a special place within environmental analysis since many of the compounds of concern contain chlorine atoms. Had it not been for the development of this highly sensitive and specific detector (for some compounds 10–100 times more sensitive than the flame ionisation detector), much of the trace analysis required for these compounds would not have been possible.

More recently, mass spectrometry has found use as a sensitive and highly selective detection method. Although the potential of this technique was apparent for many years, its widespread application had to await the development of inexpensive bench-top integrated gas chromatograph–mass spectrometer set-ups rather than the more expensive and cumbersome combination of separate instruments. Advances had also to be made in the availability of cheap computer data processing and storage to handle the massive amount of information produced even from a single chromatographic separation. This method is becoming increasingly routine. However, detailed discussion of applications is being left to the final Part 8 on ultra-trace analysis where its advantages are of greatest importance.

Columns and Stationary Phases

The range of columns available is extensive. In the choice of column for a particular application, not only does the chromatographer have to consider the most appropriate stationary phase but also the column dimensions. The column dimensions not only affect the separation efficiency, but must also be considered for compatibility with the detector used, the method of sample introduction and the sample type.

The column types available can be divided as follows, in order of decreasing separation efficiency:

Narrow-bore capillary columns:

Typical dimensions   30–60 m length, 0.2 mm i.d., flow rate
0.4 ml min^{-1} of He.

Wide-bore capillary columns:

Typical dimensions   15–30 m length, 0.53 mm i.d., flow rate
2.5 ml min^{-1} of He.

Packed columns:

Typical dimensions   2 m length, 2 mm i.d., flow rate 20 ml min^{-1}
of He.

Most recent analytical methods for water analysis use the first two types
of column, but you may occasionally find packed columns in long-
established methods or for less demanding applications.

Narrow-bore columns offer the greatest detection sensitivity and are
used for analyses close to the limits of detection. The low carrier gas
flow rate is well suited for applications where mass spectrometric
detection is used. However, direct sample injection on to the column
using a syringe is not possible as the column would become overloaded.
A splitting device is necessary for the introduction of the sample.

Wide-bore columns have a larger sample capacity and direct syringe
injection is possible. The sample may also be introduced from a sample
concentration system such as a 'purge-and-trap' device, described later
in this section and used for the analysis of volatile components. The
greater sample capacity may be required if a low-sensitivity detector is
being used. Wide-bore columns are also less affected by contamination
from non-volatile components in the sample and so find a use with highly
contaminated samples such as waste water.

Many of the organic compounds of environmental interest are of high
relative molecular mass and have low volatility. High oven temperatures
are necessary for these and consequently silicone polymers are often the
favoured stationary phases. As with other uses of gas chromatography,
the best separation efficiencies are achieved when the stationary phase
has a similar polarity to the components of the analyte. Fuel oils are
separated on non-polar columns (e.g. dimethylsilicone), pesticides and
chlorinated solvents are often separated on medium-polarity columns
(e.g. diphenyl/dimethylsilicone), whereas 2,3,7,8-tetrachlorodibenzo-*p*-
dioxin can be separated from its isomers using highly polar columns
(e.g. cyanopropylsilicone).

The stationary phase may be adsorbed or chemically bonded on to the
column walls of capillary and wide-bore columns, or on to a support

material in packed columns. For analyses close to the limit of detection and at high oven temperatures, column bleeding may become a significant factor. The use of low-loaded columns (0.1–0.25 μm film thickness) or chemically bonded phases may reduce this effect. A higher loading of columns (1–5 μm film thickness) is possible at lower temperatures for the analysis of volatile compounds. Thicker films have higher sample capacities for highly concentrated components, but there is a corresponding decrease in column efficiency compared with thinner films.

∏   How would you confirm that a peak is due to a single component rather than two components with identical retention times?

A chromatogram should be produced on two columns of different polarities. It would be unlikely that the peaks would remain unresolved on both columns.

Many standard procedures specify the use of two columns, the second column being known as the confirmation column. It is for this type of problem that capillary columns show their greatest advantage over packed columns. Their greater separation efficiency decreases the probability of unresolved peaks.

Extraction Procedures

Most analytical methods for organics involve the extraction of the compounds from water before the chromatographic analysis.

∏   The sensitivity of the available gas chromatographic detectors is in many cases high enough to allow direct injection of the water sample into the chromatograph. This is rare in practice. Can you think of reasons why?

1. Many (but by no means all) gas chromatographic columns are incompatible with water.

2. Extraction techniques can be used which are selective towards the analyte and will simplify the subsequent chromatogram.

3. Direct injection of the sample would deposit non-volatile solids on the column, which would cause blockage and shorten the column lifetime.

4. Concentration techniques lower the achievable detection limits.

The extraction methods used are common techniques in chemical analysis. Try to think of methods you have already come across before studying the following list. Figure 4.2b summarises the methods.

a. Solvent extraction. The water sample is shaken with an immiscible organic solvent in which the components are soluble. Hexane and light petroleum are the most common extraction solvents but oxygenated and

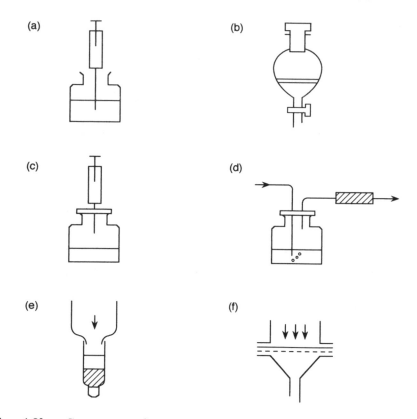

Fig. 4.2b. *Summary of extraction methods.* (a) Direct injection; (b) solvent extraction; (c) headspace analysis; (d) purge and trap; (e) solid phase extraction; (f) extraction filters

chlorinated solvents are sometimes used. The organic layer is separated and, after drying, is injected into the chromatograph. The extractions can be made selective towards acidic and basic components by altering the pH of the aqueous layer. If the sample is acidified, the basic components are less likely to be extracted. For example:

$$RNH_2 + HCl \longrightarrow RNH_3^+Cl^-$$

amine,   amine hydrochloride,
soluble in   less soluble in
non-polar   non-polar
solvents   solvents

Similarly, if the sample is made basic, acidic components are less likely to be extracted. For example:

$$RCO_2H + NaOH \longrightarrow RCO_2^-Na^+$$

carboxylic   carboxylate salt,
acid,   lower solubility
soluble   in non-polar
in non-polar   solvents
solvents

When making the choice of extraction solvent, the response of the chromatographic detector should always be considered. Hexane or light petroleum will appear as the predominant peak in the subsequent chromatogram if flame ionisation detection is used. The least interference will be caused if the solvent peak appears before the analyte peaks but there is still a potential problem with peaks from trace impurities in the solvent. Because of this, even analytical-grade solvents may have to be redistilled prior to use.

If an electron-capture or other selective detector is used, it is possible to use an extraction solvent for which the detector has low sensitivity. Hexane and light petroleum would both be suitable.

If an unsuitable solvent cannot be avoided (e.g. if a chlorinated or oxygenated solvent is required with subsequent electron-capture detection) and the analyte has low volatility, it is possible to evaporate the extract to dryness and dissolve the residue in a compatible solvent. It is better, however, to avoid this if at all possible.

b. Headspace analysis. The water sample is placed in a container with a septum seal in the lid and an air space above the sample. The simplest procedure is then, after allowing for the air to equilibrate with the water, to inject an air sample (containing volatile organic components) into the chromatograph.

This technique overcomes problems found in the previous method arising from solvent interference. The sensitivity towards a particular component will, however, be dependent on its volatility, favouring low molecular mass, neutral components.

The sensitivity of the technique may be increased by heating the sample. Beware, however, that you are also increasing the vapour pressure of the water. Care should be taken to check the water compatibility of the chromatographic column.

c. Purge and trap techniques. These techniques extract the volatile organic content from the sample by using a purge gas stream. In many instruments the organics are collected in a short tube of absorbent material such as activated charcoal or a porous polymer (e.g. Tenax). After the collection period the tube is flash heated to release the organics into the chromatograph. Other instruments collect the volatile components into a secondary liquid nitrogen cold trap. Rapid heating of this trap releases the organics into the chromatograph.

d. Solid-phase extraction. Applications of this technique are rapidly increasing. The water sample is passed through a short disposable column containing 100–500 mg of adsorbent material. The column packing is reversed-phase material, similar to that used in highb performance liquid chromatography. An ODS packing, which contains octadecylsilane groups chemically bonded on to a silica support, is commonly used. The organic compounds are retained on the extraction column and are then eluted with a suitable organic solvent (e.g. hexane). By suitable choice of extraction column material, selective retention and elution of the analyte can be achieved, providing both sample clean-up and analyte concentration. The only pretreatment necessary is to pass a small quantity of conditioning solvent through the column prior to use.

A further development is the use of extraction discs where the absorbent material is held within a filter disc. The extraction procedure is simply to pass the sample by suction through the filter. The extracted components are then eluted using a suitable solvent.

Analysis of Extractable Components

With a suitable choice of analytical column, the above extraction methods may be sufficient pretreatment for the direct injection of the extract into the chromatograph. For instance, the UK HMSO method for halomethanes simply uses the extraction of the compounds into light petroleum and injection of the extract directly into the chromatograph. The chromatogram of such extractions may be complex (particularly if flame ionisation detection is being used), a single extraction stage usually having insufficient specificity to simplify the chromatogram greatly. Indeed, a simple extraction with injection of the extract into the chromatogram is often used as a survey method to identify organic compounds in water.

Analysis of Individual Trace Pollutants

For analysis of individual components expected to be found at low concentration (e.g. pesticides), further pretreatment may be necessary. A general pretreatment scheme could be

1. extraction;

2. clean-up to remove interfering components;

3. concentration of extract.

Until the continuing development of solid-phase extractions, solvent extraction was the most often used technique for the first stage. The low volatility of many of the compounds of interest in this category renders the alternative vapour-phase extraction methods difficult.

The clean-up stages will invariably be chromatographic, most frequently using column chromatography. This may involve more than one separation stage.

To illustrate the method it is easiest to study one analysis in detail. For this purpose, an analytical scheme for the commercial pesticide DDT has been chosen.

Analysis of DDT

DDT was the first synthetic insecticide to come into widespread use. It was introduced after the Second World War, and although now controlled or banned in many areas of the world (particularly in the West), it is a universal contaminant. In common with most commercial products, the insecticide is not a single chemical compound, the major active component (p,p'-DDT) only being 70–80% of the total content. One of the minor components, p,p'-DDD (similar in structure to p,p'-DDT with a $CHCl_2$ side-chain rather than CCl_3) is, in fact, more toxic to insects than p,p'-DDT.

When considering environmental samples, a number of decomposition and metabolic products will also be present. Some of the reactions producing these materials have been considered in Section 2.3. In fact, for many samples, the highest concentration component is not p,p'-DDT but its primary metabolic product, DDE.

The following chromatographic peaks are expected in DDT analysis:

(i) Components of technical DDT:

p,p'-DDT (70–80%);

o,p'-DDT (15–20%);

p,p'-DDD (1–4%).

(ii) Decomposition products:

p,p'-DDE (aerobic decomposition);

p,p'-DDD (anaerobic decomposition).

We have a multi-component mixture even without the presence of any other compounds expected in the water sample! Interfering components in a typical sample could include other pesticides and polychlorinated biphenyls. These often have similar extraction properties to the DDT components.

A typical pretreatment would be as follows.

1. Extraction of the organic components into hexane. A 40-fold concentration increase can be achieved by extraction of a 2 l sample into 50 ml solvent.

2. Drying the solvent using a column containing 5 g of sodium sulphate. The chromatographic columns in the subsequent stages of the procedure are deactivated by the presence of water, so drying the extract at the earliest possible stage is essential.

3. Further concentration of the extract to 1 ml by partial evaporation on a steam-bath and further reduction in volume by bubbling dry nitrogen or air through the warmed sample. The sample is then placed on the top of the first chromatographic column described below.

4. Clean-up of the extract by column chromatography.

   (a) Alumina–silver nitrate column.

   Alumina is a polar column material and will retain polar components in the extract. The silver nitrate helps to retain compounds containing unsaturated carbon–carbon bonds. Non-polar material, including the DDT components, is eluted using 30 ml of hexane. The extract is then reduced in volume to 1 ml and added to the top of the second column.

   (b) Silica gel column.

   This is a less polar column than the first and can be used to separate potential non-polar interferences from the sample. Hexane (10 ml) is passed through the column, eluting polychlorinated biphenyls (PCBs), with the DDT components being retained on the column. DDT is then eluted with a more polar solvent mixture (12 ml of 10% diethyl ether in hexane).

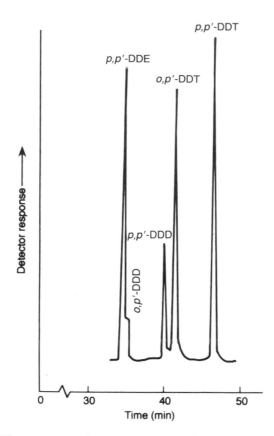

Fig. 4.2c. *Chromatographic separation of DDT components. 25 m ×*
0.32 mm i.d. methylsilicone capillary column with a temperature gradient
to 220 °C

The eluates are then re-concentrated to 1 ml (what is the overall
concentration factor during the pretreatment?) before injection into the
chromatograph. A typical chromatogram is shown in Fig. 4.2c. The
detection limit for each component is approximately 10 ng l^{-1}.

The above procedure is just one method of pretreatment. Other chromatographic methods may be used, such as preparative-scale thin-layer
chromatography or solid-phase extraction. Each of these methods will
still, however, be made up of the same individual stages of extraction,
concentration and removal of selected interfering components.

Later parts of the Unit will extend the use of this method to the analysis of solids (Part 4) and to ultra-trace components (Part 8).

∏   Clean-up of the extract simplifies the subsequent chromatogram. Can you think of a second advantage?

This is simply protection of the column and detector from contamination. Without clean-up the column lifetime will be shortened and the detector sensitivity lowered. Cleaning detectors to restore the sensitivity is very time consuming.

Fingerprinting Oil Spills

If a film of oil is discovered in water, the first question is:

'What is it? Petrol? Fuel oil? Paraffin?'

The next question may be:

'Where did it come from?'

The commercial products mentioned are complex mixtures of organic compounds. The precise composition of the mixture can vary from sample to sample, and so even if a complete quantitative analysis of every component in the mixture was undertaken there would still be much difficulty in interpretation of the data.

A simpler procedure is to produce a chromatogram under standard conditions (column packing, flow rate, column temperature) and to compare the trace either with a library of reference materials, or preferably with the results for a sample of the material suspected to have been discharged. Capillary columns are necessary for the resolution of individual components. Often correspondence of retention times and the overall envelope shape of the chromatogram will be sufficient to characterise the effluent. Hydrocarbon fuels give chromatograms with regularly spaced peaks (consecutive members of homologous series of compounds within the fuel). Lubricating oils have fewer resolved peaks. Natural product (vegetable) oils have simpler chromatograms with few individual peaks. Further information can be obtained if individual components in the material can be identified. Thus the presence of simple polyaromatic species such as anthracene will identify coke oven fractions.

Sample preparation from water containing low concentrations of hydrocarbons involves simply extracting the material with a suitable volatile solvent (e.g. diethyl ether). After washing and drying, the extract is concentrated using a stream of dry nitrogen.

With heavily polluted water, the organic material is separated by extractive distillation with toluene. The oil can then be recovered by fractional distillation from the toluene.

Chromatograms of typical commercial oils are shown in Fig. 4.2d.

Complications can occur in the interpretation of the chromatogram for a material which has not been sampled immediately after discharge. The composition of oil spills change with time. Volatile components of the oil evaporate, in general lower molecular mass components disappearing first. The oil will also slowly be biodegraded, the rate of degradation of a particular component being dependent on its chemical structure. A straight-chain hydrocarbon, for example, will be degraded more quickly than its branched-chain isomer.

What apparently seems a very simple method of identification of a pollutant is in fact a skilled task needing a great deal of experience.

4.2.3. Liquid Chromatography

The most common form of high performance liquid chromatography uses ultraviolet absorption as its method of detection. Few pollutants have sufficient ultraviolet absorption for direct detection and so analytes would have to be derivatised before analysis. There would seem to be few advantages over gas chromatography for trace organics in water samples.

Although less commonly used, alternative detectors are, however, available for HPLC and these may have advantages for specific classes of compounds.

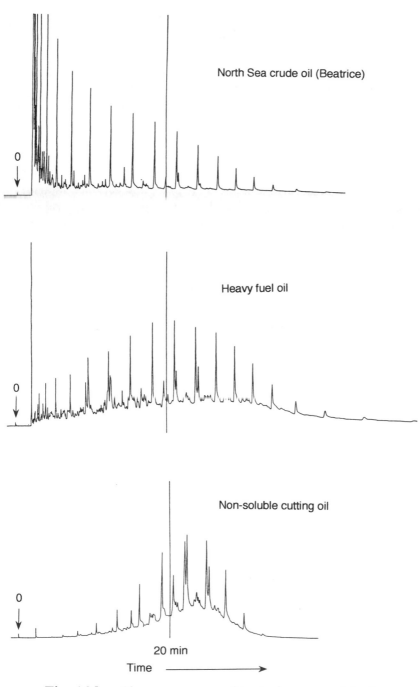

North Sea crude oil (Beatrice)

0

Heavy fuel oil

0

Non-soluble cutting oil

0

20 min

Time

Fig. 4.2d. *Chromatograms of typical commercial oils*

∏   From your knowledge of liquid chromatography, what alterna-
      tive detection techniques may find use in environmental analy-
      sis?

Conductivity detection can be used for ionic species (Section 3.4.3).

Fluorescence detection has an extremely high sensitivity and selectivity
to specific groups of compounds and may find use for these species.

Conductivity detection has found widespread use for inorganic ions.
Low molecular mass carboxylic acids (e.g. formic and acetic acid) have
very similar physical properties to the inorganic acids and ion chromatography provides a convenient alternative to gas chromatography for
these acids.

One group for which fluorescence detection has high sensitivity are
polynuclear aromatic hydrocarbons (PAHs). Examples are shown in Fig.
4.2e. They are highly carcinogenic compounds produced in trace quantities whenever fossil fuels are burnt. Typical water extracts could include
up to 70 PAHs with a total concentration of around $1 \, \mu g \, l^{-1}$.

In order to monitor these low concentrations, sample preconcentration
is needed. Solid-phase extraction using an octadecylsilane (ODS) column, or a combination of ODS and amino columns, has been used.
Sensitivity can be maximised if the detector is capable of changing the
excitation and detection wavelengths throughout the chromatographic
run, since each component has different optimum settings. The range of
wavelengths used is 270–300 nm for excitation and 330–500 nm for
detection.

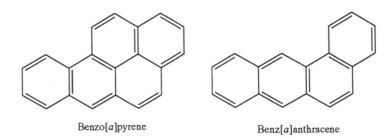

Benzo[a]pyrene   Benz[a]anthracene

Fig. 4.2e. *Typical PAHs found in environmental samples*

Fluorescent derivatives can be made from non-fluorescent or weakly fluorescent compounds. Phenols and *N*-methylcarbamate pesticides are often analysed in this way.

The procedure for *N*-methylcarbamates uses post-column derivatisation. The HPLC eluate is hydrolysed with sodium hydroxide at 95 °C, producing methylamine. The methylamine is then reacted with *o*-phthalaldehyde and 2-mercaptoethanol to produce the fluorescent derivative. The fluorescent excitation wavelength is 230 nm and detection is carried out at >418 nm, giving a limit of detection of approximately $1\ \mu g\ l^{-1}$ per component for a 400 μl sample, injected without pre-concentration.

HPLC with ultraviolet detection is sometimes used. *N*-Methylcarbamate, urea and triazine pesticides can be analysed by this method. These are 'second generation' pesticides which have been developed to replace organic halogen compounds. Detection limits with UV detection are higher than those with fluorescence detection and preconcentration (solvent extraction or solid-phase extraction) has to be applied prior to injection. This form of detection is also less specific than fluorescence and there is a greater possibility of chromatographic interference from other components in the sample.

As with phenols, the development of liquid chromatographic methods often stems from difficulties encountered with analyses using gas chromatographic techniques. In many cases this may be attributed to the polarity of the molecules (e.g. phenols, *N*-methylcarbamates) or their thermal lability (e.g. *N*-methylcarbamates, phenylureas).

4.2.4. Spectrometric Methods

Often a technique is required to measure the total concentration of groups of compounds, rather than individual concentrations. Such determinations include the analysis of:

total phenols;
surfactants (total, anionic, cationic and non-ionic surfactants);

total hydrocarbons.

Visible spectrophotometry is often used for phenols and surfactant analysis after the formation of derivatives. Chromatographic methods have in the past not been used as they give too much information!

The simplicity of the method can be seen by the analysis of anionic surfactants using a methylene blue method. Under basic conditions a salt is formed between the methylene blue and the surfactant and this salt can be extracted into chloroform. The absorbance of the extract is measured in the visible region (at 652 nm) and the concentration determined by comparison with a calibration graph.

Infrared spectrometry is used for the total hydrocarbon content. The hydrocarbons are extracted from the acidified water using a non-hydrocarbon solvent (e.g. carbon tetrachloride) and the absorption is measured at $2920 \, cm^{-1}$, corresponding to the C—H stretching frequency.

SAQ 4.2a

The Environmental Protection Agency (USA) lists 114 organic priority pollutants and suggests a purge-and-trap technique for volatile components and solvent extraction techniques for non-volatiles. Solvent extraction is under either acidic or basic/neutral conditions.

Which technique could be used for the following compounds?:
1. Toluene
2. Anthracene
3. 2,4,6-Trichlorophenol
4. Methylene chloride
5. Chloroform
6. 1,2-Dichlorobenzene
7. Phenol
8. Naphthalene
9. Hexachlorobenzene
10. Benzene

SAQ 4.2a

SAQ 4.2b What GC column would be your initial choice for the analysis of:
(*a*) chlorinated pesticides in a natural water sample;
(*b*) volatile solvents in waste water;
(*c*) oil contamination in water?

4.3.   METAL IONS

In this section we shall be looking at the analysis of metal ions found
in the $\mu g \, l^{-1}$ to $mg \, l^{-1}$ concentration range. The only metals likely to
be found above this range are the four ions already discussed in previous
sections (sodium, potassium, calcium, magnesium). Of the remaining
metals, iron, manganese and zinc can sometimes reach towards the
$mg \, l^{-1}$ level, but other metal ions, if present, are likely to be at the lower
end of this range.

Metal ions can occur naturally from leaching of ore deposits and also
from anthropogenic (man-made) sources. Such sources include metal
refining, industrial effluents and solid waste disposal. Much solid waste,
including power station fly ash, sewage sludge and harbour dredgings,
contains significant concentrations of metal ions (up to $1000 \, mg \, kg^{-1}$
total metal) which can leach into solution if in contact with water.

This area of analysis is currently dominated by techniques which can
be grouped together under the general title of atomic spectrometry. The
individual techniques are:

Flame atomic absorption spectrometry (Flame AAS);

Flameless atomic absorption spectrometry (Flameless AAS);

Inductively coupled plasma optical emission spectrometry (ICP-OES);

Inductively coupled plasma mass spectrometry (ICP-MS).

These will be discussed along with other methods, showing the relative
merits of each technique and their potential application.

4.3.1.   Storage of Samples for Metal Ion Analysis

You should by now be able to decide upon suitable sample containers
and storage conditions applicable to most metals.

∏   What are these?

Polyethylene bottles are less likely to contaminate the sample with metal ions than glass bottles.

The only exception to the use of polyethylene bottles is for mercury analysis when glass bottles should be used. Mercury ions readily react with many organic materials.

The sample should be acidified to minimise precipitation of metal ions. A typical procedure is the addition of 2 ml of 5 mol l^{-1} hydrochloric acid per litre of sample.

Scrupulous cleaning of bottles is important. This usually includes an acid washing stage to ensure complete removal of trace metals. For aluminium, the concern over contamination extends to glassware used in the subsequent analysis. You are often advised to pre-leach glassware with dilute nitric acid and to reserve glassware solely for aluminium determinations. This procedure would in fact be good practice for all metal analyses.

4.3.2. Pretreatment

Most routine analyses require the total metal content of the sample regardless of its chemical nature. The response of atomic spectrometric methods is largely independent of the chemical form of the metal. Most of the other techniques only respond quantitatively to free metal ions. Pretreatment can include evaporation to dryness and dissolution of the residue in acid, partial evaporation with acid or digestion with acid at an elevated temperature for several hours. This is to dissolve suspended material and ensure the metal is present as the free ion.

The more modern techniques we shall be discussing (flameless AAS, ICP-OES, ICP-MS) are sufficiently sensitive and interference free for most samples require no further pretreatment. Most of the other analytical techniques require an extraction/concentration step for trace analyses. Such a step can also serve to remove potentially interfering ions which may be present in far greater concentrations than the analyte. The most common method used is the formation of a neutral complex with an ion and extraction of this into an organic solvent (simple metal salts or ionic complexes would not extract). Up to a 20-fold increase in concentration is possible in a single stage. The complexing agent used depends on the subsequent analytical procedure and this will be discussed in the relevant sections.

Other extraction/concentration methods include the use of chelating or ion-exchange columns. The metal ions are first held on the column, either by complex formation with the column packing material (chelating column) or by ion exchange. The ions are then eluted as a concentrated extract with an appropriate solvent, often an aqueous buffer.

Solid-phase extraction methods are currently being developed as an alternative to these techniques. Once again the formation of neutral complexes (e.g. thiocarbamate complexes) is necessary for the extraction.

4.3.3.   Analytical Methods

4.3.3.1.   Atomic Spectrometry

We shall start with a discussion of the technique you are probably most familiar with—flame atomic absorption spectrometry—and then show how the other atomic spectrometric techniques overcome problems found in its use for trace metal analysis.

∏   From your knowledge of the technique, draw and label a diagram of a flame atomic absorption spectrometer.

A diagram of a flame atomic absorption spectrometer is shown in Fig. 4.3a.

A light beam of the correct wavelength to be specific to a particular metal is directed through a flame. The flame atomises the sample, producing atoms in their ground (lowest) electronic energy state. These are capable of absorbing radiation from the lamp.

Although the equipment appears completely different from other forms of absorption spectrometry, the law by which the absorption of light is related to concentration is similar to that we have used already for absorption of visible and ultraviolet radiation. *What is this?*

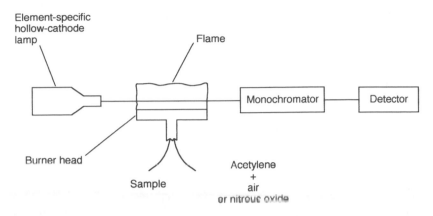

Fig. 4.3a. *Diagram of a flame atomic absorption spectrometer*

This is the Beer–Lambert law. If you cannot write out the law in mathematical form, revise Section 3.4.1.

The concentration range over which the law applies for atomic absorption spectrometry is usually $0-5 \text{ mg l}^{-1}$.

Over the last two decades atomic absorption spectrometry has dominated routine analyses of metal ions in aqueous samples.

∏   From your knowledge of atomic absorption spectrometry, can you think of some of the advantages of the technique?

It is a rapid technique and can easily be automated.

It is a simple method for routine use.

Standard procedures are available for all metals.

The analyses are generally free from interferences, and known interferences can easily be overcome.

Apart from the pretreatment stages already mentioned, little or no sample preparation is needed for aqueous environmental samples.

You may have included 'high sensitivity' within your list. However, this has been left out above as it is discussed further below.

Atomic absorption spectrometry is indeed sensitive and, if it is used for the more common ions discussed in the Section 3.4, the water samples would have to be diluted before analysis. Magnesium is analysed by flame AAS, often after sample dilution. If the technique is used for sodium or potassium analysis, lower sensitivity absorption lines, rather than the highest sensitivity lines, would be used in addition to diluting the sample. Atomic emission (flame photometry) is, however, the preferred technique for these ions.

If flame AAS is used for trace metal analysis, preconcentration of the sample is necessary. This may simply involve partial evaporation of the acidified sample for zinc, iron and manganese analyses. Solvent extraction has been routinely used for other metals.

Since atomic absorption analysis is relatively free from interference from other trace metal ions (i.e. the presence of other materials usually has little effect on the accuracy of the analysis) the extraction need not be highly specific to any one particular metal. In fact, it may be beneficial to be able to use a single complexing agent for several metals since the extraction stage is the most time-consuming part of the analytical procedure. Ammonium pyrrolidinedithiocarbamate (APDC) (Fig. 4.3b) is often used as it forms stable complexes with most transition metals, if the pH is correctly adjusted.

After extraction of the analyte into an organic phase, the latter is aspirated directly into the flame. The increase in the sensitivity is above

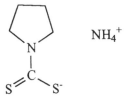

Fig. 4.3b. *Ammonium pyrrolidinedithiocarbamate (ammonium tetra-methylenedithiocarbamate)*

that which is expected from the simple concentration factor. This is due to the increased aspiration rate resulting from the lower viscosity of the organic solvent in comparison with water.

You should be able to see a number of disadvantages of solvent extraction/flame atomic absorption:

— it is very time consuming;

— the sensitivity may still be insufficient for low-concentration metal ions;

— the risk of sample contamination is considerably increased.

To overcome these problems other atomic spectrometric techniques have been applied to trace metal analysis.

Flameless atomic absorption spectrometry. By replacement of the flame by other methods of atomising the sample, the sensitivity can be increased sufficiently to remove the need for sample preconcentration. For most metals this would mean the use of graphite furnace atomisation but, as we shall see later, it is not the only method possible. A comparison of flame and graphite furnace AAS is summarised below.

Advantages of solvent extraction/flame AAS	Advantages of graphite furnace AAS
1. Simple technique	1. Increased sensitivity (μg l^{-1} concentrations).
2. The solvent extraction stage can be used to remove potential interferences.	2. Decreased overall analytical time as the solvent extraction stage is not usually necessary.
3. More readily available equipment.	3. Smaller samples required.
4. Shorter instrument time.	4. Unattended operation is possible.
5. Lower instrument cost.	5. Reduced risk of sample contamination.

As you can see, the chief advantage of flameless AAS arises from removing the necessity for preconcentration of the sample. An extraction stage may still sometimes be necessary for complex samples in order to reduce potential interferences as is the case for sea water analysis.

One major source of error is background interference due to light scattering by solid particles within the beam. The scattering is highly dependent on wavelength:

$$\text{scattering} \propto 1/\lambda^4$$

where λ = wavelength of radiation.

The analytical wavelengths used for lead and cadmium are towards the far end of the available ultraviolet range and so analyses for these elements are highly susceptible to interference. Automatic background correction should always be used for these elements. An analytical wavelength of 283.3 nm is also often preferred for lead rather than the more sensitive 217 nm wavelength, and this lessens the effect of light scattering.

Other flameless atomisation techniques can be used for specific elements. Inorganic mercury salts can be chemically reduced using tin(II) chloride or sodium tetrahydroborate. The elemental mercury produced is then swept by a stream of nitrogen or air into a gas cuvette for absorption measurement in a modified spectrometer.

Tin, lead and a number of metalloids (As, Se) can be reduced by sodium tetrahydroborate to volatile hydrides, which are swept from the sample by a gas stream. Mild heating breaks down the hydrides to produce the element in its ground state.

Inductively coupled plasma optical emission spectrometry (ICP-OES) and induct-ively coupled plasma mass spectrometry (ICP-MS). Atomic absorption spectrometry has a number of disadvantages for use in analysing large numbers of samples of varying elemental composition and concentration.

∏   What are the two major problems in the use of AAS for these samples?

AAS can only determine one element at any one time. The technique becomes slow and tedious for multi-element analysis. The variations in concentrations of the samples can be problematic as the linear range of AAS is very limited.

The development of ICP techniques for water analysis can be seen as an attempt to overcome these problems. At the same time they maintain the advantages of graphite furnace AAS of being sufficiently sensitive not to require a preconcentration stage and also in not using flammable or explosive gases. This permits unattended, 24 hour operation.

In both methods the sample is atomised in a plasma flame at 6000–10 000 K. With ICP-OES, the emission spectrum is monitored. Simultaneous ICP-OES can determine 60 or more elements at once by monitoring at preset wavelengths. Sequential spectrometers, which are more common for water analysis, are restricted to a smaller number of elements, determined by the requirements of the analysis, measured in succession by rapid changes in the detection wavelength. The total analysis time is still fast, typically 5 s per element.

In common with other emission techniques, there is the problem of spectral overlap from different elements, an element producing many more lines in its emission spectrum than in its corresponding absorption spectrum. The choice of the analytical wavelength is based on freedom from interference and the sensitivity. For routine water analysis this problem has been largely overcome, sensitive and interference-free lines being well documented.

The sensitivity of the technique is very element dependent but is generally between those of flame and graphite furnace AAS. This makes ICP-OES useful for taking over the analyses previously performed using flame AAS but only the higher concentration analytes measured by furnace AAS. Its wide linear range (approximately 10^5) is a particular advantage for water analysis, meaning that trace metals can be measured simultaneously with higher concentration species.

A more recent development is to use the inductively coupled plasma as an ion source for a mass spectrometer. We shall be dealing with mass spectrometry in more detail when used as a detector for gas chromatography in Part 8. It is sufficient here to remind you that charged ions (in this case produced by the plasma flame) are separated according to their mass/charge ratios. The mass spectra of inorganic mixtures are simple in comparison with the more familiar organic compound spectra and few interferences occur in metal analysis. Although the technique is not strictly simultaneous, the ions being determined sequentially, determination of 20 elements is possible within 4 s. The sensitivity is slightly lower than that of graphite furnace AAS but is still sufficient to determine trace metal ions at below 1 μg l^{-1} in aqueous samples.

Each development in atomic spectrometry has brought with it a significant increase in instrument capital cost. The cost of instrumentation is generally flame AAS < furnace AAS < ICP-OES < ICP-MS. In addition, ICP techniques have significantly higher running costs owing to the consumption of argon necessary to generate the plasma. The advantages of ICP techniques are, however, so great that ICP-OES has for several years been the preferred technique for the high-concentration metal analyses in major water analysis laboratories, with ICP-MS rapidly becoming established for lower concentration metals. Atomic absorption methods find a role in smaller laboratories where the sample throughput is insufficient to justify the additional capital and running costs of ICP techniques.

4.3.3.2. Visible Spectrometry

Until the widespread use of atomic spectrometric techniques, visible spectrophotometry was the most commonly used technique for metal ion analysis. Standard methods were developed for all commonly found metal ions. These methods use colour-forming complexing agents. Selectivity in the analysis is achieved in two different ways:

1. Solvent extraction is sometimes used. Chromium is analysed as the diphenylcarbazide complex after extraction into a trioctylamine/chloroform mixture. This gives a limit of detection of 5 μg l^{-1} in the original sample. The complexing agent dithizone can be used for 17 metals. The selectivity is achieved by precise control of pH and the use of masking agents.

2. Alternatively, a colour-forming complexing agent can be used which is sufficiently sensitive and selective for use in the aqueous sample without extraction being necessary. Examples are shown below.

Metal	Reagent	Limit of detection ($\mu g\ l^{-1}$)
Iron(II)	2,4,6-Tripyridyl-1,3,4-triazine	60
Manganese	Formaldoxime	5
Aluminium	Pyrocatechol violet	13

A number of these techniques have been adapted for use with portable colorimeters (e.g. iron, manganese, chromium, copper) and it is perhaps in this area that the techniques have the most widespread current usage.

It is useful to consider why such a well established technique as visible spectrometry could become largely superseded by atomic methods:

1. Atomic methods are more rapid.

2. Although visible spectrometric pretreatment is generally simple when analysing relatively unpolluted water samples (rivers, lakes), they may become complex and time consuming with more complicated samples such as sewage effluents.

3. Visible spectrometry is often affected by interference from other elements. This can be illustrated by the determination of iron using 2,4,6-tripyridyl-1,3,5-triazine. The following concentration effects were observed on a true value of $1.000\ mg\ l^{-1}$ iron:

Additional ion	Effect on measured ion concentration
100 mg l^{-1} sulphate	$-0.020\ mg\ l^{-1}$
2 mg l^{-1} cadmium	$+0.009\ mg\ l^{-1}$
10 mg l^{-1} lead	$-0.026\ mg\ l^{-1}$

Nonetheless, visible spectrometry remains a frequently used technique and would be the method of choice when atomic spectrometric methods are unavailable.

4.3.3.3.   Anodic Stripping Voltammetry

A number of electrochemical methods are sufficiently sensitive to determine the low levels of metal ions typically found in the environmental water samples without separate preconcentration. Anodic stripping voltammetry has found particular use in environmental analysis where at least 19 metals can be analysed.

The apparatus consists of an electrolytic cell containing a working electrode (a mercury drop, or a thin film of mercury deposited on a glassy carbon electrode), a reference electrode and a counter electrode. A three-electrode system is used so that current and applied potential can be measured independently of each other. This attempts to compensate for the change in potential drop due to the resistance of the test solution during the analysis which would affect the measurement in a two-electrode system.

The sample is placed in the cell along with a supporting electrolyte (e.g. $0.1 \, mol \, l^{-1}$ acetate buffer at pH 4.5). Nitrogen or argon is bubbled through the solution to remove dissolved oxygen, which would otherwise interfere in the analysis. The working electron is held at a small negative potential with respect to the reference whilst the solution is stirred. Reduction of the metal ions to the free metal occurs at the working electrode:

$$M^{2+} + 2e \longrightarrow M$$

Under controlled conditions of deposition time and stirring rate, the quantity of metal deposited on the electrode is proportional to its original concentration.

After a predetermined time, the potential of the electrode is slowly changed in the positive direction. At specific potentials, depending on the metal and supporting electrolyte, each metal is oxidised and returned back into solution:

$$M \longrightarrow M^{2+} + 2e$$

This process is monitored by plotting the current change between the working and counter electrodes against the potential (Fig. 4.3c). The

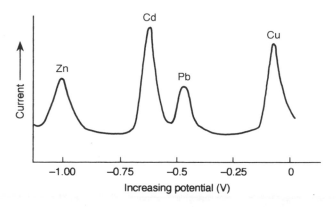

Fig. 4.3c. *Typical anodic stripping voltammogram*

height of the peak in the curve is proportional to the concentration of the metal.

∏   Can you see why the method does not require a separate concentration stage?

If you consider the experimental method, you will see that the first step, where the metal is being plated on the electrode, is itself a concentration stage, hence the high sensitivity of the technique with little pretreatment.

The supporting electrolytes necessary for individual metals are tabulated in standard texts. Electrolytes are often acidic, as potentials become little affected by minor changes in the sample composition. Complexing agents (e.g. acetate) are sometimes included to stabilise particular oxidation states or to move the stripping potential of the metal away from potential interferences. Four metals of major environmental concern, copper, lead, cadmium and zinc, can, however, be analysed in a single scan using the acetate buffer mentioned earlier.

Anodic stripping voltammetry does have at least one disadvantage. The method is slow (stripping times can vary from 30 s to 30 min), during which time the apparatus is devoted to a single sample. Compare this with atomic spectrometry where the instrumental time is only a few seconds per sample.

As an electrochemical method, the technique determines only free metal ions in solution and to some extent loosely associated complexes. If the total metal content is required, sample pretreatment is necessary. This may range from simple acidification to UV irradiation to destroy any potential complexing agents. By performing the analysis with and without pretreatment, a measure of the free and complexed metal ions can be made, which would not be possible by atomic spectrometry. This makes anodic stripping voltammetry a useful research tool, but because of its relative slowness limited in application for routine analysis.

4.3.3.4. Liquid Chromatography

As environmental waters invariably contain a large number of metal ions, and often at similar concentrations, you might think that liquid chromatography would be a frequently used analytical technique. Your argument might be that it should be possible to determine all metal ions present using a single sample injection into the chromatograph. Atomic methods still, however, dominate metal analysis. Liquid chromatography only finds use in areas where atomic spectrometry is not ideal.

∏   Re-read the section on atomic absorption spectrometric analysis of metals and decide areas where liquid chromatography may be preferable to this form of atomic analysis.

This is not the easiest of questions, but you may have found some of the following:

1. Complex matrices.

   Extraction techniques are often necessary when complex samples are analysed by AAS in order to remove interfering components. This extends the time taken to perform an analysis considerably.

2. Analysis of mixtures of uncommon elements.

   AAS determines individual elements using a different lamp for each element. Additional, and perhaps unsuspected, elements will not be detected. With the correct choice of column and eluent these would be seen as additional peaks in a liquid chromatographic analysis. The need to change lamps for each element may also mean that AAS is a slower technique than chromatography for complex mixtures. You may also find that it is more difficult to obtain hollow-cathode lamps necessary for AAS for the more uncommon elements.

3. Quantification of different chemical forms of the ion.

Later we shall be discussing the different chemical forms in which a metal can be found in the environment. In certain instances, ion chromatography can separate and quantify the chemical forms. Atomic absorption spectrometry is unable to distinguish the different species

If we extend our comparison to atomic emission techniques, then many of the perceived advantages may still hold. Interferences in complex samples may still be found. Sequential plasma emission spectrometers detect a limited number of elements and unsuspected elements may still be missed. Different chemical forms are not distinguished. Emission techniques are, however, more rapid for multi-element analyses.

Chromatographic methods using both dedicated ion chromatographs and conventional HPLC have been developed. The most sensitive method for transition metals in complex mixtures using a dedicated chromatograph is known as 'chelation ion chromatography.' This method involves the use of two preconcentration columns (Fig. 4.3d) and spectrometric detection after mixing with a derivatisation agent, 4-(2-pyridylazo)resorcinol. Detection limits are 0.2–1 μg l^{-1} with a 20 ml sample volume.

After acid digestion to ensure the metal present is as the free ion, the sample is added to the first column, eluted on to the second column and subsequently on to the analytical column. Although we can say that ion chromatography removes the need for a *separate* preconcentration stage, you can see that preconcentration still occurs within the instrument as an identifiable analytical step.

The separation of Fe^{3+}, Cu^{2+}, Ni^{2+}, Zn^{2+} and Mn^{2+} requires approximately 12 min (Fig. 4.3e), whereas separation of transition metals and lanthanides together requires 35 min.

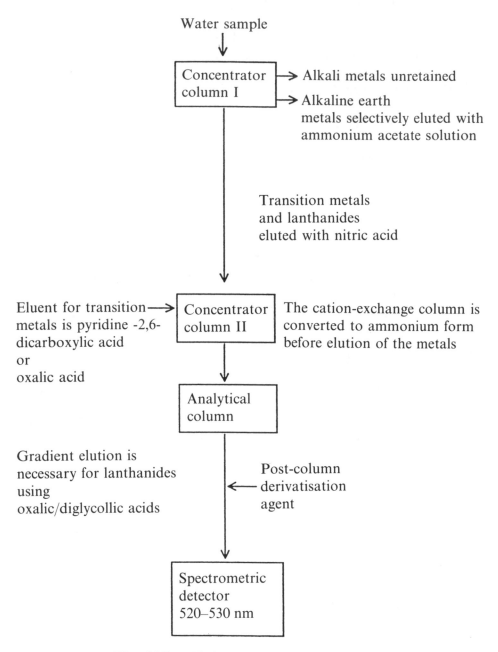

Fig. 4.3d. *Chelation ion chromatography*

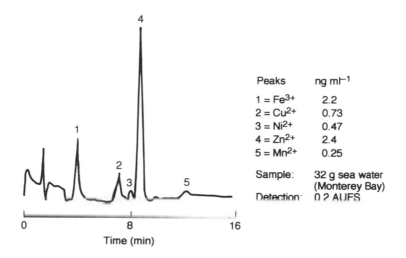

Fig. 4.3e. *A typical chelation ion chromatogram*

A typical application of chromatography to separate different chemical forms of the element is for Cr^{3+} and $Cr_2O_7^{2-}$. Once again the species are determined by visible spectrometry after derivatisation. Pyridinedicarboxylic acid is reacted with the sample prior to the analytical separation to complex with the Cr^{3+} to form a stable anionic complex (i.e. precolumn derivatisation). Separation of the species uses an anion-exchange column. Diphenylcarbazide is added after elution from the column (post-column derivatisation) to react with the $Cr_2O_7^{2-}$. Both species may be detected at 520 nm.

A common method for trace metals using conventional HPLC is by separation of the thiocarbamate complexes. Excess thiocarbamate (diethylthiocarbamate and pyrrolidinedithiocarbamate salts have both been used) is added to the sample and the transition metal complexes formed are concentrated by liquid–liquid or solid-phase extraction. The concentrated extract is injected into the HPLC system and separated using a reversed-phase technique.

Underivatised metal ions can be separated by reversed-phase ion-pair techniques or by using a cation-exchange column. Several detection methods have been used including conductivity and post-column derivatisation with 4-(2-pyridylazo)resorcinol.

4.3.3.5.   Metal Speciation: Comparison of Techniques

Speciation is defined as the different physical and chemical forms of a substance which may exist in the environment. When considering water samples, this includes not only the truly dissolved metal ions (as free metal ions or as complexes) but also colloidal forms of the metal and any metal contained within, or adsorbed on, suspended particles.

∏   What lead species do you think might exist in a river?

Some of the species are shown in Fig. 4.3f.

Although you will probably not have thought of all the species, it is hoped that you will now appreciate the great diversity of species which may be found. They include not only well defined ions and compounds, but also loosely bonded complexes and adsorbed species. The free metal ion is often only a small percentage of the total content.

	Examples	
Free metal	Pb^{2+}	
Ion pair	$PbHCO_3^+$	solution
Complexes with organic pollutants	Pb^{2+}/EDTA	
Complexes with natural acids	Pb^{2+}/fulvic acid	suspension
Ion adsorbed onto colloids	Pb^{2+}/Fe(OH)$_3$	colloidal
Metal within decomposing organic material	Pb in organic solids	
Ionic solids	Pb^{2+} held with the clay structure. $PbCO_3$	solid

Fig. 4.3f.   *Typical chemical forms of lead in a river*

The interconversion between species is slow and for many purposes the species can be considered as distinct chemical forms. The transport of each in the environment will be different and each will have different toxicological properties. As an example, let us consider the behaviour of metals within a stream and associated sediment. Any decaying vegetation will increase the metal loading in the stream water, since the organic acids formed as part of the decay process form soluble coordination complexes with the metals. The toxicity of the stream water, however, may not be increased as much as you might expect. As a very general rule, metal complexes have a lower toxicity than the free metal ion.

If the metal has more than one stable oxidation state in water, there may also even be differences in behaviour between different oxidation states. Chromium in the form of $Cr_2O_7^{2-}$ has a greater toxicity than Cr^{3+}. It would appear that the $Cr_2O_7^{2-}$ ion can enter cells via routes which permit entry of the similarly sized SO_4^{2-} ion. Such a route would not be possible for the positively charged Cr^{3+} ion.

Each of the analytical techniques that has been described in this text will respond in a different manner to the species in solution.

Technique	Response
Atomic spectrometry	All the metal species in the sample, i.e. total metal determined.
Visible absorption spectrometry	Free metal ions plus ions released from complexes by the colour-forming reagent.
Anodic stripping voltammetry	Free metal ions analysed plus any ions released from complexes during analysis. The total is often referred to as 'total ASV labile content.'
Chromatography	Non-labile species can sometimes be determined separately.

If one of these techniques is included as part of a more lengthy analytical procedure, the pretreatment stages may alter the species being analysed. Any filtration, for instance, will remove particulate matter.

Speciation may be investigated by taking advantage of the different responses of the analytical techniques and the effect of pretreatment. The most common method is to perform several ASV analyses with different pretreatment stages. A simple two-step procedure would be to perform the analysis on samples with and without ultraviolet irradiation, giving a value for the free metal (or, more precisely, total ASV labile content) and total metal content, respectively.

The complete chemical characterisation of a sample would be exceedingly complex and time consuming. When you remember that the total metal concentration may not be greater than a few $\mu g \, l^{-1}$, you realise that you may also be reaching the detection limits of the available techniques. This aspect of environmental trace metal analysis is currently of great interest and improved techniques are continually being reported in the literature.

∏   A complete characterisation of all species in a sample is a difficult and time-consuming procedure. Can you think of an alternative, and simpler, approach to species analysis to support investigations on metal transport and toxicology.

Rather than attempting to determine each species individually, species with similar environmental transport or toxicological properties could be analysed as groups. A simple classification of metal species would be into organic solvent soluble (neutral complexes, organometallic species) and organic solvent insoluble (charged complexes, free ions) species. The first type would be transported in the environment and would accumulate in fatty tissues in a similar manner to neutral organic molecules (Section 2.3) and the second type in a similar fashion to other ions (Section 2.4) within the environment.

SAQ 4.3a   Lengthy pretreatment techniques are often necessary with the analytical techniques described for both organic compounds and metals. Filtration, solvent extraction and chromatographic pretreatment are common.
What could affect the precision of measurement for low concentrations of common pollutants?

SAQ 4.3a

SAQ 4.3b
> An analytical technique for copper, lead, cadmium and zinc in water as described in the chemical literature involved dividing the filtered sample into five aliquots. The treatment of the aliquots (prior to ASV analysis) was as follows:
> 1. strong chemical oxidation and UV irradiation;
> 2. no pretreatment;
> 3. weak chemical oxidation;
> 4. passage through a chelating resin;
> 5. extraction using an organic solvent and UV irradiation of the aqueous phase.
>
> What species are determined in each aliquot?

SAQ 4.3b

Summary

Components present at trace (μg l^{-1}) levels can have a major effect on water quality if they can bioaccumulate in organisms or have a high degree of toxicity. These components usually fall into the two categories of organic pollutants and metal ions. Instrumental methods for the determination of the components have been discussed along with the necessary extraction and pretreatment steps. The predominant instrumental technique for organic components is gas chromatography, whereas atomic spectrometric techniques are the most frequently used methods for metal ion analysis.

Objectives

You should now be able to

- understand the necessity for pretreatment for analysis of water components at trace concentrations;

- choose and apply suitable methods, including pretreatment, for the analysis of organic trace pollutants in water;

- determine and describe appropriate methods for trace metal analysis in water samples;

- understand what is meant by the term 'speciation' and describe how it may be investigated for metals in water by the analytical techniques discussed.

SAQs AND RESPONSES FOR PART FOUR

SAQ 4.2a

> The Environmental Protection Agency (USA) lists 114 organic priority pollutants and suggests a purge-and-trap technique for volatile components and solvent extraction techniques for non-volatiles. Solvent extraction is under either acidic or basic/neutral conditions.
>
> Which technique could be used for the following compounds?:
> 1. Toluene
> 2. Anthracene
> 3. 2,4,6-Trichlorophenol
> 4. Methylene chloride
> 5. Chloroform
> 6. 1,2-Dichlorobenzene
> 7. Phenol
> 8. Naphthalene
> 9. Hexachlorobenzene
> 10. Benzene

Response

First let us decide which are the volatile compounds.

Toluene, methylene chloride, chloroform and benzene (1, 4, 5, 10). These will be analysed by the purge-and-trap technique.

There are two phenols in the list, *2,4,6-trichlorophenol and phenol(3, 7).* These would need to be extracted under acidic conditions (under basic conditions they would be in the form of non-extractable salts).

The other compounds would form the base/neutral group: *anthracene, 1,2-dichlorobenzene, naphthalene and hexachlorobenzene (2, 6, 8, 9).*

Do not be too concerned if you do not have all ten correct, but it is hoped that you were able to think your way through most of them.

SAQ 4.2b

> What GC column would be your initial choice for the analysis of:
> (*a*) chlorinated pesticides in a natural water sample;
> (*b*) volatile solvents in waste water;
> (*c*) oil contamination in water?

Response

(*a*) The pesticides will probably be present at the lower end of the trace level range. Even after extraction and clean-up a large number of compounds may still be present in the sample. A narrow-bore capillary column would give the required high resolution and detection sensitivity. A medium-polarity silicone polymer column would be a good initial choice.

(*b*) The components of interest would probably be part of a much higher concentration of waste chemicals. A wide-bore capillary column would be better suited than a narrow-bore column since it has a higher sample loading capacity. It would also be more tolerant of any non-volatile impurities in the sample. The column could also be coupled directly to the purge-and-trap system which you would almost certainly have used in the sample preparation. A medium-polarity silicone polymer column would again be a good initial choice.

(*c*) There would be a strong likelihood of non-volatile residues in the oil extract. A wide-bore capillary column would be more tolerant of contamination than narrow-bore columns. A non-polar silicone polymer column would be a good initial choice.

SAQ 4.3a

> Lengthy pretreatment techniques are often necessary with the analytical techniques described for both organic compounds and metals. Filtration, solvent extraction and chromatographic pretreatment are common.
> What could affect the precision of measurement for low concentrations of common pollutants?

Response

Loss of analyte at each stage is possible. This will be a particular problem if it is present in low concentration. For the determination of common pollutants, contamination of the sample may also occur. The problem increases with the number of stages of pretreatment and the number of reagents involved. Common metal ions are universally found at low concentration in all reagent solids and traces of pesticides are common in organic solvents. Low molecular mass organic materials (e.g. solvents) are themselves in use within laboratories. All materials in contact with the sample should be regarded as potential sources of contamination.

Had the question simply been restricted to metals, you might also have included the problem of different metal species in the sample. Unless you had converted the metal species completely into a single form, you would risk the loss of metal at each stage of the analysis according to the chemical behaviour of the species.

SAQ 4.3b	An analytical technique for copper, lead, cadmium and zinc in water as described in the chemical literature involved dividing the filtered sample into five aliquots. The treatment of the aliquots (prior to ASV analysis) was as follows: 1. strong chemical oxidation and UV irradiation; 2. no pretreatment; 3. weak chemical oxidation; 4. passage through a chelating resin; 5. extraction using an organic solvent and UV irradiation of the aqueous phase. What species are determined in each aliquot?

Response

The pretreatment in (1) ensures that all the metal species in the sample will have been decomposed and so the *total metal* concentration will be determined.

Procedure (2) will measure the free metal ions and loosely bound complexes. This measurement is known as *ASV labile metal.*

Procedure (3) will destroy easily oxidisable organic material. Subtraction of the results, (3) − (2), will measure *metal found in complexes.* This is known as *organically bound labile metal.*

Free metal ions will be held by the resin in procedure (4), as will the metal from loosely bound complexes.

Procedure (4) will then determine *metal bound in highly stable complexes.*

Procedure (5) will extract organic soluble complexes, hence (1) − (5) will measure the *organic-soluble* content.

5. Analysis of Solids

OVERVIEW

This part of the book introduces you to methods for the sampling and extraction of solids. These are necessary stages prior to completion of analysis by the instrumental techniques as already discussed for water samples.

5.1. INTRODUCTION

Up to the present we have had an easy time with the complexity of our analytical schemes. The analytes were already largely dissolved in water, and required minimum treatment for the subsequent stages of the chemical analysis. If we now extend our scope to include the analysis of solids, we shall have to consider an extra stage, that is, the extraction of the species of interest. This may in fact be the most difficult part of the analysis. The current part of the book is devoted to this preliminary stage, along with problems encountered with the sampling of solids.

Before you start this part of the book, you should be sure that you can remember the principles of the transportation of pollutants discussed in Part 2 as these are necessary to understand the relevance of the analyses.

∏   What solids do you consider to be of importance for the study of the environment? What specific analyses would be relevant?

1. Animal and plant specimens

These are directly of interest since the toxic effect of a compound is proportional to its concentration within the organism.

The investigations would also be relevant to species further along the food chain to determine the environmental pathway of the pollutant (Section 2.3).

Plants and animals may also be used as indicator organisms to monitor pollutants found in lower concentrations in the wider environment. As an example, heavy metal pollution in sea water is often monitored by analysis of seaweed rather than by direct analysis of the water. Remember, however, that you have to balance the advantage of the higher concentration in the living organism with the disadvantage of the more complex analytical matrix.

The effect of pollution on living organisms can sometimes also be investigated by monitoring levels of naturally occurring constituents of the organism. The effects of acid rain on trees, for instance, include a decrease in concentration of alkaline earth ions in the leaves.

2. Soils

Soils are complex materials comprising of

weathered rock;

humus;

water;

air.

They provide nutrients for plants, in addition to providing physical anchorage and support for growth. Nutrients include nitrogen (in the form of nitrate and ammonia), phosphorus (in the form of orthophosphate) and trace metals such as copper, iron, manganese and zinc.

Not all the ionic material within soil can, however, be extracted by plants. Some is too strongly bound within the soil structure. Although a total analysis of soil is sometimes performed, the more frequency need is the determination of *available* ionic material.

The transport of material in the soil is influenced by the acidity or alkalinity of water in the soil structure, so soil pH is frequently monitored.

A further common analysis is for metal ion or organic contamination of the soil, resulting, for instance, from the misuse of pesticides, dumping of waste material or deposition of pollutants from the atmosphere.

3. Sediments and sewage sludge

We have already discussed how both organic compounds with low water solubility and metal ions tend to accumulate by adsorption on freshwater or marine sediments (Sections 2.3 and 2.4). Analysis of the higher concentrations found in sediments may be an easier task than analysis of the surrounding water. Analysis may be of the adsorbed species only, or may be of the total sediment. The sediment is often fractionated according to particle size prior to the analysis. This is important as the adsorption of pollutants can often be related to the available surface area, which in turn is related to particle size. The particle size also affects the mobility of the sediment and the possibility of ingestion by marine organisms.

Sewage sludge is the inert material produced as the end product of the sewage treatment process. The material is sometimes spread on land as a soil conditioner or may be disposed of by incineration or dumped as a waste product. The greatest concern over this material is the metal content, which may be as high as $1000 \, \text{mg kg}^{-1}$ total metal in some sludges.

4. Atmospheric particulates

An important route for the transport of inorganic salts and neutral organic compounds is via atmospheric particulate deposition (Section

2.2). Typical determinations are elemental analysis of particulate material or analysis the organics adsorbed on the particulate surface. Once again particle size may have to be determined.

5.2. COMMON PROBLEM AREAS IN THE ANALYSIS OF SOLIDS

In the following sections we shall discuss the analysis of plant and animal specimens, soils, sediments and sewage sludges, reserving discussion of particulates until we have looked at other components of the atmosphere. Despite the diverse nature of these solids, their analyses have common problem areas. After making a number of general comments on sampling, pretreatment, extraction and analytical determination common to all the solids, we shall look at the analysis of each solid in detail.

5.2.1. Sampling

Concentrations of the analyte may vary widely from sample to sample, both on a local scale (e.g. adjacent soil samples taken from a field) and on a larger scale (adjacent fields). 'Identical' plants can possess different levels of contamination. For example, leaves on the windward side of trees are exposed to atmospheric pollution to a greater extent than those on the leeward side. The variation, or inhomogeneity, has to be reflected in the number of samples to be taken. The subsequent analysis could be of each individual sample, or after combining a large number of samples and subsampling to obtain an average concentration.

Sampling positions have to be chosen with care. For monitoring exercises involving large areas, the area should be subdivided using grids and ideally the sampling points should be at fixed locations within each grid. Invariably other considerations (e.g. geographic features or the availability of suitable plant specimens) prevent this. Areas where there is the possibility of specific contamination from other sources should be avoided. It would be easy, for example, to sample soil for general pollution by monitoring close to roads, because of ease of access and proximity to the ideal grid location, whilst at the same time forgetting the localised pollution from the road traffic. Whenever possible, duplicate samples should be taken at each location to assess local inhomogeneity.

If monitoring is related to a point source into the atmosphere (e.g. factory emissions) then consideration should be given to the prevailing wind direction with an increase in the number of sampling positions in the area of highest likely contamination. Prevailing currents should be taken into account for marine discharges.

Control samples should also be taken at points remote from the area under investigation and an effort should be made to match the control site as closely as possible to the sample site. If, for instance, a factory discharge located in an urban environment is being monitored, then the control site should be an urban site where there is no possibility of similar contamination.

5.2.2. Pretreatment

This can include:

washing of sample;

drying;

grinding/homogenisation.

These procedures are often deceptively simple. It is easy to forget that most samples are biologically active and even washing, prolonged warming or storage at room temperature may change their composition. In addition, some analytes may be thermally unstable, volatile or even photolytically unstable. Contamination or analyte loss is also possible at each stage of the analytical procedure.

5.2.3. Extraction of the Analyte

This could involve any of the following processes depending on the analysis.

solvent extraction—neutral organic compounds;

ashing and subsequent dissolution—elemental composition;

extraction using aqueous solutions—'available' ions.

Many of the analytes are very common contaminants (some, such as DDT, are classed as 'universal' contaminants) and may be present in the extraction agents or adsorbed on the apparatus. High-purity reagents specific to the analysis should be used. For instance, 'pesticide-free'-grade solvents are available from some manufacturers. A blank sample should be included in the analytical scheme to monitor contamination during the analytical process, but the preparation of a 'blank' sample may in itself be difficult for universal or near-universal contaminants.

5.2.4. Analytical Determination

Instrumental methods of analysis generally follow the procedures in Parts 3 and 4.

∏   What did we find were the most common procedures for the analysis of organic compounds and for metals in aqueous samples?

What extra considerations would you think necessary for analysis of solid extracts?

Most organic materials would be analysed by gas chromatography. Atomic absorption spectrometry is the most common simple technique for metals.

Due consideration would have to be taken of the possible interference from other components extracted. For gas chromatographic analysis, this could take the form of careful assessment of the suitability of sample clean-up technique prior to the chromatographic analysis (Section 4.2.2). High-resolution columns may be necessary. For analysis by atomic absorption spectrometry, background correction may be required, or again development of new clean-up techniques may be necessary (Section 4.3.3).

SAQ 5.2a

> Loss of analyte and contamination of the sample may occur during lengthy pretreatment stages. Suggest two methods which could be used to assess and compensate for errors introduced during pretreatment.

5.3. SPECIFIC CONSIDERATIONS FOR THE ANALYSIS OF BIOLOGICAL SAMPLES

We shall discuss the sampling and extraction of components from plant material and later consider differences in approach which may be necessary for animal tissue.

5.3.1. Sampling and Storage of Plant Material

The sample may be foliage, roots or the whole plant. A single species should be sampled with each specimen, if possible, being at a similar stage of maturity. If foliage is being sampled, the minimum sampling height should be such that there is no possibility of contamination by upward splashing from the soil, assuming that the species is tall enough! The maximum height is often determined by practical considerations. A suitable sample size is often 500–1000 g. The sample may be stored under refrigeration for a few days if it cannot be analysed immediately.

5.3.2. Pretreatment

Washing

Even a simple procedure such as washing may extract the analyte. Patting the sample dry using a paper tissue can result in contamination of the sample with trace metals. It is consequently often preferable to avoid washing altogether if suitable clean samples can be found. Soft brushing may be an alternative. The cleanliness of samples is particularly important for trace metal analysis where the concentration may be higher in the surrounding soil than in the plant specimen.

Some pollutants may have been deposited from the atmosphere on to the leaf surface. If you are studying uptake of the pollutant by the plant, then this would have to be removed by washing. If, however, you are studying transfer of the pollutant along the food chain, then this should be included or determined separately. Dioxins, for instance, are not taken up by plants, but can enter the food chain by deposition on leaves, which are then eaten by herbivores.

Drying and Homogenisation

⊓ What two factors do you consider will determine the temperature and drying time of a biological sample?

The temperature and duration of drying must be a balance between too low a temperature over a protracted period promoting biological activity and too high a temperature over a shorter time period leading to loss of volatile components.

A typical drying procedure would be to blow a current of dry air over the sample for a period of up to 12 h. The temperature should not be in excess of 50 °C. Alternatively, the sample may be freeze-dried, involving deep-freezing the sample, reducing the pressure and removing the water by sublimation.

⊓ Owing to the risk of potential losses caused by drying, why do you think it is necessary at all?

Your answer will probably be so that the analytical result can be referred to a dry mass. There is, however, no reason why you could not calculate the dry mass on a second sample. Drying the samples lessens the possibility of change due to biological activity. A second advantage is that homogenisation of the bulk sample, necessary if sub-samples are to be taken, is made easier if the sample is dry.

Homogenisation of the dried sample often is by use of a high-speed grinding mill. Care should be taken, once again, to ensure that you are not introducing contaminants in the grinding process.

5.3.3. Extraction Techniques for Organic Contaminants

If present, organic contaminants are likely to be in the $\mu g\ kg^{-1}$ concentration range or below. The simplest method for the extraction of organics is to shake a sub-sample with an extracting solvent (e.g. hexane or light petroleum for neutral organics) and to leave the two phases in contact for several hours.

An alternative method is to use Soxhlet extraction. The apparatus is shown in Fig. 5.3a.

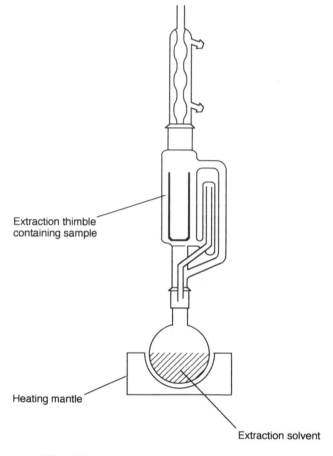

Fig. 5.3a. *Soxhlet extraction apparatus*

Fresh solvent is continuously refluxed through the finely divided sample
contained in a porous thimble, and a syphon system removes the extract
back into the refluxing solvent. The net effect is continuous extraction
by fresh solvent. A typical extraction takes 12 h. The technique is only
applicable to analytes which can withstand the reflux temperature of the
solvent.

∏   The use of hexane or light petroleum as extraction solvent
     assumes a dried sample. What problems could you foresee if for
     any reason the sample was not dried?

     Suggest methods by which this could be overcome.

The two solvents suggested are immiscible with water and would easily form an emulsion. In some cases it may be difficult for the solvent to penetrate the sample.

A desiccant is often mixed with the sample during the extraction. Sodium sulphate is commonly used. The solvent can be modified by the addition of a polar solvent such as acetone. Alternatively, the extraction solvent can be changed completely to a solvent which is miscible with water. Acetonitrile is often used. You must, of course, be certain that the solvent is still appropriate for the extraction of the analyte.

If the sample will not allow the solvent to penetrate the structure, conditioning the sample with a polar solvent which is miscible with both water and the extraction solvent may overcome the problem. Propan-2-ol has been used for this purpose.

5.3.4. Ashing and Dissolution Techniques for Trace Metals

Trace metals are likely to be in the mg kg^{-1} concentration range. The concentrations will vary from species to species and throughout the growing season.

For the extraction of metals the organic matter is decomposed by dry or wet ashing. Dry ashing consists of heating the sample in a muffle furnace, typically at 400–600 °C for 12–15 h. The resulting ash is then dissolved in dilute acid to give a solution of the metal ions. Inaccuracies can arise both from volatilisation of metals and the retention of metals in an insoluble form in the crucible.

Wet ashing consists of heating the sample with oxidising agents to break down the organic matter. A typical procedure would be heating with concentrated nitric acid followed by perchloric acid. Alternative combinations include sulphuric acid/hydrogen peroxide and nitric/sulphuric acids.

An advantage of wet digestion is lower losses from volatilisation (due to lower temperatures and liquid conditions), but it can give rise to higher metal blanks from impurities in the acids. Great care has to be taken with methods using perchloric acid. Perchloric acid in the presence of carbonaceous material or metals has a tendency to detonate on drying! Small sample sizes should always be used and the liquid in the flask should never be allowed to dry out.

5.3.5. Analysis of Animal Tissues

Although the above discussion is specifically for plant analysis, much is applicable to the analysis of animal tissue if certain differences in the sample types are recognised. Animal samples are more liable to decomposition than plant samples and should be preserved by freezing below $0\,°C$. Organic compounds are extracted without drying the sample. Often the bulk sample is homogenised in a blending mill with water and sub-samples are taken from the slurry for extraction. Remember also the previously described method of extraction from moist samples by inclusion of a solid drying agent.

An alkaline digestion stage is also often included before organic extraction to break down any fatty tissue.

Metals are once again extracted after wet or dry ashing. Typical concentrations in specific tissues for both metal ions and neutral organic compounds can be in the $mg\,kg^{-1}$ range.

SAQ 5.3a

A method sometimes suggested to ensure complete extraction of the organic pollutants from biological samples is to repeat the extraction a number of times with different solvents. What disadvantage would this technique have for the subsequent stages of the analysis?

SAQ 5.3a

5.4. SPECIFIC CONSIDERATIONS FOR THE ANALYSIS OF SOILS

5.4.1. Sampling and Storage

Soil composition may vary greatly over a small area. We have already discussed how samples will have to be taken from a number of locations to obtain a suitable average composition. There will also be differences in composition according to the depth of sampling.

For soil contamination studies you should take into account the source of the contamination and its mobility within the soil. Often pollutants deposited from the atmosphere are immobile and will remain within the surface layer; lead contamination from vehicle exhausts decreases rapidly (within a few centimetres) with depth. Dioxins similarly remain in the top layer of soil, the molecules becoming strongly bound within the soil structure.

If the soil is disturbed (e.g. by ploughing), samples should be taken from the whole of the disturbed area. If the investigation is concerned with possible uptake by plants or crops, then the sampling should be over the whole depth that the root system penetrates (which may be greater than the depth of any ploughing). For landfill sites samples should be taken over the complete depth of the landfill.

Typical samplers are shown in Fig. 5.4a.

Samples should be cooled or frozen for transportation to the laboratory.

∏   Suggest reasons why there will be changes in the composition of soils throughout the year.

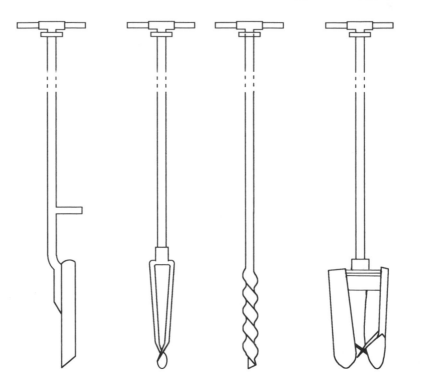

Fig. 5.4a.   *Typical soil samplers*

You may have included the following:

biological activity, which will consume nutrients, will be greatest in spring and summer;

rainfall, which may leach out components, will vary throughout the year;

pesticides and fertilisers will only be applied at specific times in the year.

5.4.2.   Pretreatment

Drying

Until you remember the great amount of microbial activity in a typical soil sample, it may not seem obvious why as much care has to be taken with drying soil samples as with biological samples. Soil is often dried by equilibrating with the atmosphere at room temperature (under certain circumstances this may be raised to 30 °C) for not less than 24 h. Under harsher conditions the levels of available nutrients may change (this is particularly the case for phosphorus, potassium, sulphur and manganese) and nitrogen-containing compounds interconvert. The last problem is so great that analyses for nitrogen compounds should always use field-moist samples.

Grinding

The drying process leaves the soil in large aggregates which need to be broken down into the constituent particles whose sizes range from 2000 μm for the coarse sand component to less than 2 μm for clay. This is done using a mortar and pestle, after sieving to 2 mm to remove pebbles and other large particles.

Subsampling

This is always a problem with solids, as any agitation tends to fractionate mixtures according to particle size. The smaller particles tend to fall below the larger particles. Standard methods are well established to overcome this problem, the simplest being the cone and quartering technique. The total sample is formed into a symmetric cone. The cone is then divided vertically into segments and alternate quarters are combined, the remaining half being rejected. The process can be repeated successively until the required sub-sample size is produced.

5.4.3. Extraction of Organic Contaminants

Organic contamination is typically in the μg kg^{-1} concentration range. Extraction closely follows the techniques used for plant samples in Section 5.3.3.

5.4.4. Extraction of Available Ions

Concentrations of available trace metals and available nutrients would be expected to be in the mg kg^{-1} range. We shall divide our discussion into a general method for the extraction of ions and then look at additional procedures which are needed for the special case of the measurement of nitrogen availability.

First we need to know what is meant by the term 'available.'

The complex structure of the soil acts as an ion exchanger for both cations and anions, where the simple ions are held to the soil by ionic forces. These ions may only be released into solution from the soil if the total charge remains constant. This release will be dependent on the soil type and the chemical composition of the extraction water. Analytical procedures for available ions attempt to reproduce the environmental conditions by suitable choice of extracting solution.

The procedure is simply to shake the soil with the extracting solution for a fixed period, typically 1 h. A range of extracts have been used, including ammonium acetate solution, dilute acetic acid, dilute hydrochloric acid and EDTA solution, to mimic local conditions. This has led to problems in comparison of results with other laboratories where a different extractant may have been used and consequently a different proportion of the ions released.

Once in solution, most ions can be analysed by the methods discussed in Parts 3 and 4.

Nitrogen Availability

Nitrogen species found in soil are:

organic nitrogen;

nitrate;

nitrite;

ammonia (free ammonia and ammonium ion).

Only the last three constitute the readily available nitrogen, although the organic nitrogen content is subject to microbial decay, which will release nutrients over a period of time. A measurement of organic nitrogen should thus also be included in the scheme.

The ionic forms of nitrogen can be extracted with potassium chloride solution. A subsequent reduction with, for example, titanium(III) sulphate quantitatively converts all the ions into ammonia, which can then be determined by standard methods.

∏   What is the standard method for ammonia analysis?

This is by increasing the pH of the solution with sodium hydroxide, distilling the ammonia into water and titrating with standard acid (Section 3.4.4).

Organic nitrogen is measured after a preliminary conversion into ammonia. This is achieved by boiling with concentrated sulphuric acid for several hours (Kjeldahl method). Potassium sulphate is added to raise the boiling point of the sulphuric acid, along with a catalyst (selenium or mercury is often used). Typical apparatus is shown in Fig. 5.4b.

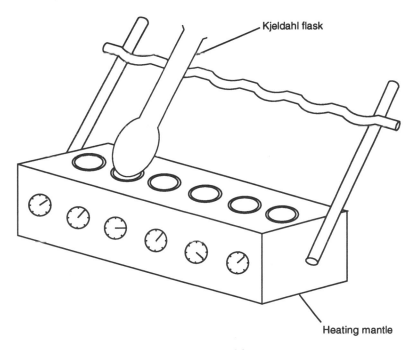

Fig. 5.4b.   *Kjeldahl apparatus*

5.4.5.   Dissolution Techniques for the Determination of Total Metal Concentrations in Soil

The *available* metal concentration as determined in the last section is only part of the *total* metal concentration in the soil. The total concentration analysis is occasionally required for environmental investigations. Extreme conditions have to be used to dissolve the soil such as dissolution in hydrogen fluoride/perchloric acid mixtures or fusion with an alkaline flux (e.g. sodium carbonate) and subsequent dissolution in dilute acid.

Once in solution, the metal concentration can be determined by the standard techniques described in Parts 3 and 4.

5.4.6.   Determination of pH

Although soil contains water as an essential constituent, it is of course predominantly a solid. Since pH can only be defined as hydrogen ion concentration in solution, then the pH of soil is the pH of water in equilibrium with that soil.

∏   This definition gives a hint to a difficulty in this seemingly simple analysis. What is the problem?

The water is in equilibrium with the soil. Any change of the conditions (even adding more water) can alter the equilibrium and hence the pH.

At the very least a thick paste of soil and water is necessary to measure the pH. The added water should be such that there is minimum disturbance to the solution equilibria. A salt solution (potassium or calcium chloride) is often used to form the paste, and this is left for 1 h for the equilibria to be re-established.

SAQ 5.4a
> A monitoring exercise is planned for lead deposited on soil close to a busy roadway. How would you select sampling positions?

SAQ 5.4a

SAQ 5.4b What instrumental methods would be used for the analysis of the following in soil extracts?:

Potassium

Calcium

Magnesium

Available phosphorus (present as orthophosphate)

Trace metals

5.5. SPECIFIC CONSIDERATIONS FOR THE ANALYSIS OF SEDIMENTS

5.5.1. Sampling and Storage

The first problem with sediment analysis is to obtain the sample from the river or sea bed. Core samplers are available for shallow areas (Fig. 5.5a). The tube is immersed with the valve system open. The valve is then closed to permit the sample to be withdrawn. Just before breaking the surface of the water, the tube is sealed to preserve the sediment structure so that sections corresponding to different depths in the sediment can be analysed. This can provide a historical record of the deposition of pollutants.

Grab samplers may be used for greater depths, or where the sediment is loose so that there is no vertical structure.

Samples are often stored deep frozen. The concentration of organic pollutants is often in the μg kg^{-1} range.

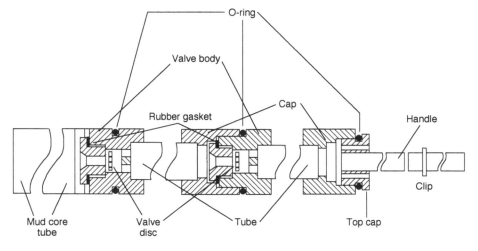

Fig. 5.5a. *Core sampler*

5.5.2. Pretreatment

∏   What is the main difference between sediment samples and other
     samples we have looked at, which may modify the pretreatment?

It is the high water content. On return to the laboratory, the sample is
thawed and screened to remove large contaminants such as stones and
twigs, if necessary using pressure. In most of the previous methods we
have looked at the sample is then dried. It would be impracticable to
remove the water from sediment by air drying at room temperature and
if the analysis is to determine organic material, the analysis proceeds
using wet samples. Samples for metal ion analysis may be oven dried at
110 °C before further treatment.

Pretreatment may also include the separation of the sample size fractions
by wet sieving.

∏   A map showing the distribution of trace metals around a marine
     site for the dumping of solid waste showed a wider geographic
     distribution of metals in the lower size fractions. What is happen-
     ing?

It is simply that the smaller particles are drifting more in the local
currents.

Can you remember the other reasons why analysis of size fractions may
be useful? Look at Section 5.1 again if you are unable to do this.

5.5.3. Extraction Techniques for Organic Contaminants

Organic analysis is once again based on solvent extraction of the
homogenised slurry (produced using a high-speed blender), often using
a Soxhlet apparatus. The extraction solvents we have previously used
are often extremely hydrophobic. If they were used with a wet sediment
sample we would end up with water–solvent emulsions. In order to
overcome this, more polar extractants such as acetonitrile or acetone
are used.

5.5.4. Dissolution Techniques for Trace Metals

Acid dissolution is used for the analysis of adsorbed metal ions, but care has to be taken not to dissolve the bulk sediment itself. A suitable acid mixture would be concentrated nitric acid/hydrogen peroxide. Concentrations in the $mg\,kg^{-1}$ range could be expected in the sediment. Extreme solubilisation techniques have to be used for analysis of the less soluble portion of the sediment. A typical method uses hydrogen fluoride under pressure in a Teflon-lined Parr bomb. We shall be looking at other methods for insoluble solid analysis in Part 7.

∏ Why do you consider that in some cases analysis of adsorbed metal ions is most relevant for a study of environment problems, whereas in other cases analysis of the whole of the sediment is more appropriate?

Environmental analysis is concerned with ions and compounds which are available to living species. Only loosely adsorbed metal ions will always be readily available. The availability of the ions in the bulk of the sediment will depend on, amongst other criteria, the particle size and the chemical composition.

5.5.5. Analysis of Sewage Sludge

This is included in this section owing to the high water content of the sample, which requires similar pretreatment for trace level organics to those discussed above for sediment samples. The material however, has a high organic content and so digestion is necessary before any metal analysis. Typically this would involve heating with concentrated nitric acid in a Kjeldahl apparatus (Fig. 5.4b) and extraction of the metal ions after dilution with water.

SAQ 5.5a

What are the relative merits of investigating metal or insoluble organic compound pollution in a river or sea by:
(a) analysis of the water;
(b) analysis of the sediment;
(c) analysis of the fish or shellfish?

SAQ 5.5a

Summary

Most chemical analytical techniques rely on the analyte being present in solution. This part of the book examines extraction and dissolution techniques from solids to solubilise the components of interest. The analysis can then proceed by the instrumental techniques which have already been discussed. Solids which are of importance in studying the environment include animal and plant specimens, soils, sediments and sewage sludge and atmospheric particulates. Specific extraction and dissolution procedures have been discussed for each type of solid except atmospheric particulates. Atmospheric particulates are dealt with in Part 7, after discussion of the gaseous components of the atmosphere.

Objectives

You should now be able to:

• understand the problems of sampling and pretreatment of solid samples prior to the analytical determination of organic compounds and metal ions;

- apply your knowledge of these problems to the analysis of:

    plants and animals;

    soil;

    sediment;

    sewage sludge.

SAQs AND RESPONSES FOR PART FIVE

SAQ 5.2a   Loss of analyte and contamination of the sample may occur during lengthy pretreatment stages. Suggest two methods which could be used to assess and compensate for errors introduced during pretreatment.

Response

One method is to add a known quantity of an internal standard to the sample as early as possible during the chemical manipulations. The behaviour of the standard must follow that of the unknown as closely as possible. However, this method does not overcome errors introduced by incomplete initial extraction from the solid.

It is also possible to include the analysis of reference material of accurately known composition in your analytical scheme. Such samples are available for a wide variety of materials from National Laboratories. Once again the reference sample should match the unknown material as closely as possible.

SAQ 5.3a

> A method sometimes suggested to ensure complete extraction of the organic pollutants from biological samples is to repeat the extraction a number of times with different solvents. What disadvantage would this technique have for the subsequent stages of the analysis?

Response

The use of a number of solvents will increase the number of extracted components. This would then require either more comprehensive clean-up procedures prior to chromatographic analysis, and/or a higher degree of resolution in the chromatographic separation. You may also have to consider problems associated with additional solvent peaks in the chromatographic analysis and the trace impurities introduced with each solvent.

SAQ 5.4a

> A monitoring exercise is planned for lead deposited on soil close to a busy roadway. How would you select sampling positions?

Response

The deposition would be expected to be greatest at the side of the road and would decrease with distance. A site should be selected where samples could be taken close to the road, ideally where there is no intervening pavement. The site should be distant from any other potential sources of lead. Both sides of the road should be sampled to compensate for effects of wind. Sampling (taking duplicate samples) should be more frequent close to the road, perhaps with sampling distances from the road in the ratio $1:2:4:8:\ldots$ Since the source of the lead is from the atmosphere, surface samples should be taken. Samples taken below the soil surface could give an indication of penetration into the soil. Reference samples as similar as possible to the monitoring samples should be taken some distance from the roadway.

SAQ 5.4b

What instrumental methods would be used for the analysis of the following in soil extracts?:

Potassium

Calcium

Magnesium

Available phosphorus (present as orthophosphate)

Trace metals

Response

Potassium and calcium can be analysed by flame photometry (see Section 3.4.2) and magnesium and the trace metals by atomic absorption spectrometry or other atomic spectrometric techniques (Section 4.3.3). Orthophosphate is best determined by visible spectrometry after conversion to a blue phosphomolybdenum complex (Section 3.4.1).

SAQ 5.5a

What are the relative merits of investigating metal or insoluble organic compound pollution in a river or sea by:
(*a*) analysis of the water;
(*b*) analysis of the sediment;
(*c*) analysis of the fish or shellfish?

Response

Water will be the easiest to sample (unless you have access to local fisheries) and will need less pretreatment to remove potential interferences. The concentrations determined will be lower than in the other samples, and often are only slightly above the lower limits of routine detection. (You may recall the typical concentrations given for DDT in Fig. 2.3d and the metal enrichment factors shown in Fig. 2.2b.)

Sediment will need more pretreatment to remove potential interferences. Concentrations will vary greatly from site to site and even sample to sample and so a large number of samples would need to be taken to obtain an average concentration. On the other hand, it is ideal for investigating localised pollution. In each case the effect of enrichment or bioaccumulation makes it easier to detect the species since they will be present at higher concentrations than in the surrounding water.

Fish are not static and so it is difficult to relate concentrations found to specific locations. There may be large variations in concentration from specimen to specimen. Shellfish are more static and measured concentrations may be more easily related to localised pollution.

6. Atmospheric Analysis— Gases

OVERVIEW

This part of the book introduces you to methods for the sampling and analysis of gaseous compounds in external and internal environments. The methods include the application of both previously studied techniques and new techniques. The new techniques are discussed in detail.

6.1. INTRODUCTION

∏ List the components of clean, dry air and give an indication of their approximate concentrations.

The gaseous constituents of the atmosphere are shown in Fig. 6.1a.

You will probably have included the major components, but many of the minor components may be a surprise to you. You may have considered these to be anthropogenic pollutants. There are few of the common gases (chlorofluorocarbons being possibly the only example) which are not found in the atmosphere from natural sources. Non-localised atmospheric pollution problems are mainly concerned with increases in concentrations of naturally occurring compounds above unpolluted clean air levels.

We have come across problems due to acid rain (which contains high concentrations of sulphur and nitrogen oxides) and global warming (increasing carbon dioxide levels being a major contributing factor) in Part 1. There is also concern over rising concentrations of methane

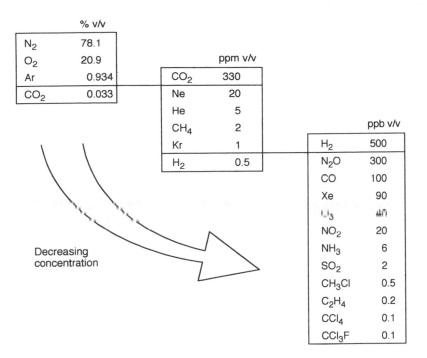

Fig. 6.1a. *Gaseous constituents of the atmosphere*

(another greenhouse gas) and of ground level concentrations of ozone (an oxidant, which produces breathing difficulties), even in non-industrialised areas.

Localised problems in urban or industrial areas can be more complex, not only due to the introduction of a large number of other pollutants but also from atmospheric reactions producing new species. A good example of this is the complex series of reactions which occur each day in large cities throughout the world. Under specific meteorological conditions (thermal inversion), pollutants build up in the atmosphere without dispersal. Gases given off by vehicles (CO, NO, NO_2, unburnt hydrocarbons) inter-react to produce a range of oxidants including ozone and peroxyacetyl nitrate (PAN). The chemical changes throughout a typical day are shown in Fig. 6.1b. These reactions produce a haze over the city, known as photochemical smog, and the compounds produced can cause respiratory problems.

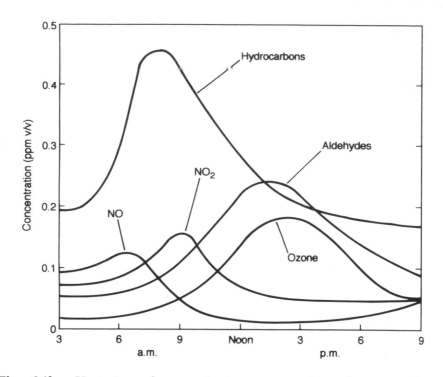

Fig. 6.1b. *Variation of atmospheric concentrations during a photo-chemical smog incident*

One of the main reasons for current environmental concern is the potential effect of pollutants (including airborne pollutants) either directly or indirectly on human health. Most of the population of the industrialised world spend their day inside buildings at work or at home. The monitoring of atmospheres within buildings (internal atmospheres) is then also of major importance. Internal atmospheres are enclosed, preventing dispersal of any pollutants. Higher concentrations of pollutant gases than are found in external atmospheres may be expected.

Internal atmospheres can also have a much wider range of pollutants than are found in external atmospheres. Large numbers of potentially hazardous chemicals are produced or used inside buildings, e.g.

gases from fuel combustion;

solvents in paints;

gases from cleaning fluids.

There are many unexpected sources. Even some forms of wall insulation can give off hazardous gases (formaldehyde).

Measurement of ambient concentrations is, however, just one of the many types of gas analysis which you may have to perform. Emission concentrations (concentrations of pollutants in flue gases or vehicle exhausts before they are dispersed into the environment) are just as important, legislation often being based on these values. Compare this with water monitoring where analysis of both discharges and the receiving water are often required.

Exhausts and flue gases can contain a much wider range of gaseous compounds than are likely to be detectable in the general atmosphere. Concentrations in the wider atmosphere may not build up to detectable levels because of the continuous removal of the gases by physical and chemical processes. Look at the list of gases which may be emitted from a typical coal-fired power station, as shown in Fig. 6.1c, and notice the discharge of hydrogen chloride, hydrogen fluoride and even mercury.

Concentrations of pollutants in atmospheres and exhaust streams can vary significantly over a short time period and, for many purposes when monitoring atmospheres, the average concentration over a period of time is required in addition to an instantaneous measurement. These are known as time-weighted average (TWA) concentrations. In the next two sections you will find that some of the analytical methods are very suited to the measurement of time-weighted concentrations, as the analyte is collected over an extended time period. One reading can then be used to give the average concentration over the whole sampling period. These methods are described in Section 6.2.

Other methods, which are used to give instantaneous concentration readings, are described in Section 6.3. In order to use these methods for

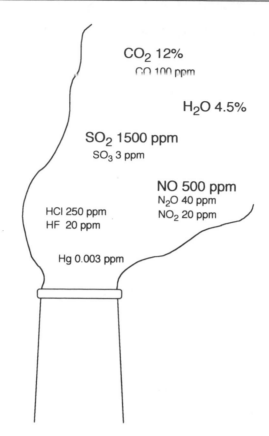

Fig. 6.1c. *Typical flue gas analysis of a coal-fired power station*

TWA measurements many readings are necessary, but this can nowadays often be readily achieved by microprocessor control of the instrument.

What, though, are the most useful time periods over which to take an average reading?

The concentration of many components in an external atmosphere varies over a 24 h cycle. Diurnal influences include the effect of sunlight and factory emissions. A concentration averaged over the 24 h cycle may be appropriate. The US National Air Quality Standard (NAQS) for sulphur dioxide (365 μg m^{-3} or 140 ppb, averaged over a 24 h period) is an example. Shorter term averages may be required for pollutants which vary rapidly throughout the day. The US (NAQS) standards for carbon monoxide are based on 8 h and 1 h periods.

Sampling times of 24 h would not be appropriate for internal atmospheres. The major concern for pollutant levels in internal atmospheres is over human health and, in particular, chronic exposure. The sampling time is then generally specified to be over an 8 h period, reflecting the length of the average working day.

You may come across a number of terms related to time-averaged exposures which are derived from national legislation. For example, the UK specifies an occupational exposure standard (OES) for most gases (Fig. 6.1d), which is defined as the concentration of an airborne substance, averaged over the reference period, at which there is no evidence that it is likely to be injurious to persons exposed by inhalation. More toxic substances, such as benzene and hydrogen cyanide, are assigned a maximum exposure limit (MEL), which is defined as the maximum time-averaged concentration to which a person can be lawfully exposed by inhalation.

Although the reference period is normally 8 h, there are also short-term (10 min) standards which apply to any period throughout the working day. This is to accommodate the possibility of acute effects from the gas. The 10 min OES for ammonia is 35 ppm or 24 mg m^{-3}.

A Note on Units

Concentrations in Fig. 6.1a are expressed as volume of analyte/total volume of sample. This is a common way of expressing gas concentrations with many direct-reading instruments calibrated in these units. We

	ppm v/v	mg m^{-3}
Ammonia	25	17
Carbon Monoxide	50	55
Methanol	200	260
Nitrobenzene	1	5

Fig. 6.1d. *Typical 8 hour average occupational exposure standards (UK) 1991*

perceive gas concentrations as volume fractions, rather than in other units. Most people would know whether an atmosphere containing 20.9% v/v oxygen would support life, but would they be so sure if it contained 9.3×10^{-3} mol l^{-1} oxygen or 299×10^3 mg m^{-3} oxygen? There is, however, one difficulty. The same units cannot be used for measuring suspended particulates. As we will see in Part 7, these are just as much of concern in environmental analysis. The alternative units based on mass of analyte/total volume may be used for both gases and particulate material, with typical concentrations being expressed as follows:

μg m^{-3} for external atmospheres;

mg m^{-3} for internal atmospheres.

Gas concentrations throughout the remainder of the book are given in both units. You will find concentrations expressed as mass/volume measurements in most national and international legislation. For every-day purposes, volume/volume measurements are also frequently used, owing to the convenient form of their expression.

The conversion between ppm and mg m^{-3} is straightforward, simply requiring the relative molecular mass of the compound, and the molar volume of the gas (for an approximate conversion this can be taken to be 24.0 l for all gases at 20 °C and 1 atm pressure). However, the units are often quoted side-by-side in tables of environmental standards. Thus the UK 8 h occupational exposure standard for toluene is expressed both as 50 ppm and 188 mg m^{-3}.

∏ The current EC monthly limit for nitrogen oxide emissions from coal-fired power stations (measured as NO_2) is 650 mg m^{-3}. What is this concentration in parts per million (volume/volume)?

Relative molecular mass of nitrogen dioxide = 46

Therefore number of moles of nitrogen dioxide in 1 m^3 air $= \dfrac{650 \times 10^{-3}}{46}$

$$= 14.1 \times 10^{-3} \text{ mol}$$

Volume occupied by 1 mol at 20 °C and 1 atm pressure = 24.0 l

$$= 0.0240 \text{ m}^3$$

Therefore, volume of nitrogen oxide in $1m^3$ air
$$= 14.1 \times 10^{-3} \times 0.0240 \text{ m}^3$$
$$= 338 \times 10^{-6} \text{ m}^3$$

Therefore, concentration of nitrogen oxide $= 338$ ppm v/v.

Look back at Fig. 6.1d for other examples of the two sets of units. Using the approximate molar volume for all gases, the conversion can be expressed as

$$\text{concentration (ppm)} = \frac{\text{concentration (mg m}^{-3})}{\text{relative molecular mass}} \times 24.0$$

The expression is identical if you convert μg m^{-3} to ppb. For compounds with molecular masses close to 24 (e.g. ammonia, carbon monoxide), concentrations expressed as μg m^{-3} and ppb are numerically roughly the same, whereas for higher molecular mass compounds (e.g. nitrogen dioxide, nitrobenzene) the numerical value of the mass/volume concentration is higher than the volume/volume concentration.

SAQ 6.1a What is the difference in meaning of the term 'parts per million' when applied to gas concentrations and aqueous concentrations?

SAQ 6.1b | Briefly summarise the expected concentration ranges of pollutants in external and internal atmospheres and in exhaust gases.
Suggest reasons why we may find there are sometimes different analytical methods used for external and internal atmospheres.

6.2. DETERMINATION OF TIME-WEIGHTED AVERAGE CONCENTRATIONS

6.2.1. Absorption Trains

This is perhaps the method which would first occur to you for monitoring trace components. A known volume of gas is bubbled through an absorbing solution. At the end of the sampling period the solution is taken back to the laboratory for analysis, generally using volumetric or spectrometric methods.

The absorption train consists of a number of containers through which the gas sample is drawn. The sample volume is measured by a gas meter, but for shorter sampling times where the flow can be kept constant the gas flow and an accurate sampling time may be used instead. The International Standards Organisation Specification is shown in Fig. 6.2a. A typical practical system would be as in Fig. 6.2b.

The reagents in the gas washing bottle are determined by the gas to be analysed. There are specific reagents for most of the inorganic gases including SO_2, Cl_2, H_2S and NH_3. Carbon monoxide is the only common exception.

A typical procedure is shown by the West and Gaeke method for sulphur dioxide, which uses a spectrometric final analysis. The sulphur dioxide is absorbed in an aqueous solution of sodium tetrachloromercurate, and the colour developed by the addition of *p*-rosaniline hydrochloride (in hydrochloric acid) and formaldehyde. The absorption is measured at 560 nm.

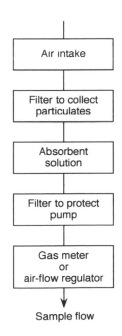

Fig. 6.2a. *Schematic diagram of absorption train*

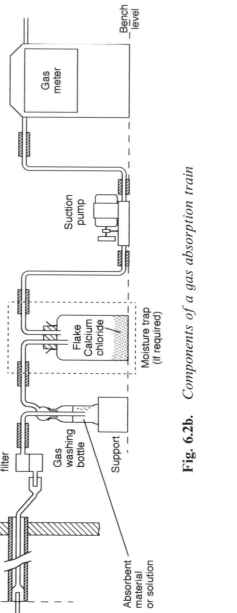

Fig. 6.2b. *Components of a gas absorption train*

The procedure above is now a reference method for the US Environmental Protection Agency (EPA), i.e. it is judged to be the best available technique and can be used to assess other methods.

∏   What properties must the absorbent possess in order to produce an accurate analysis?

1. The reagents have to be highly specific to the analyte gas.

2. The absorption of the analyte has to be quantitative. Remember that you may be analysing compounds whose concentration may only be the order of parts per billion (v/v) in the atmosphere.

3. The reagent has to be resistant to oxidation and to being stripped from solution. Remember you could be bubbling air through the solution for periods up to 24 h.

There are many advantages to this method of gas analysis. At the end of this part you will be asked to compare the methods with alternative procedures.

6.2.2. Solid Adsorbents

The most commonly used method for low-concentration organic components, particularly for internal atmospheres, is to adsorb the gas on a solid and later analyse the components by gas chromatography.

Sampling

Passive and active sampling methods can be used. Passive samplers (sometimes called diffusion samplers) consist of the adsorbent (typically activated charcoal or Tenax porous polymer) contained in a small tube sealed at one end, the other end being exposed to the atmosphere. The adsorbent is separated from the atmosphere by a diffusion zone which is either an air gap, or an inert porous polymer, according to the manufacturer. The tubes may be clipped to the lapel or carried in the breast pocket to allow personal monitoring.

An alternative design of sampler is in the form of a badge which can be clipped to the lapel. The principle of operation is similar to the tube design above.

Examples of the two types are shown in Fig. 6.2c.

Active sampling methods draw air through the sample tube by means of a pump. Sampling rates can be several hundred ml min^{-1}, but lower flow rates are often used (as low as 20 ml min^{-1}) so that sampling can continue over an 8 h period without the capacity of the tube being exceeded. Some adsorption tubes (Fig. 6.2d) contain two sections of adsorbent, the main section to be used for the analysis, the second (back-up) section to be used as confirmation that the capacity of the analytical section has not been exceeded.

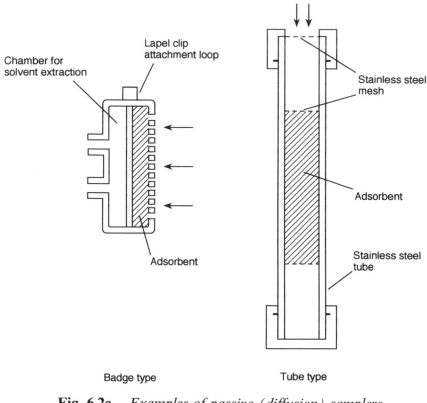

Badge type   Tube type

Fig. 6.2c.   *Examples of passive (diffusion) samplers*

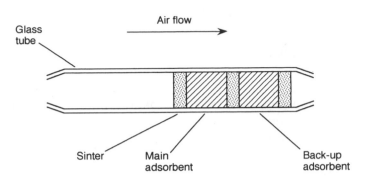

Fig. 6.2d. *A typical adsorption tube used for active sampling*

Pumps are available which are small enough to be clipped to the waist with the sample tube positioned on the lapel as shown in Fig. 6.2e. We shall encounter these 'personal samplers' again in Part 7 on particulate analysis.

The advantage of active sampling over passive sampling is that lower concentrations can be monitored for a given sampling time.

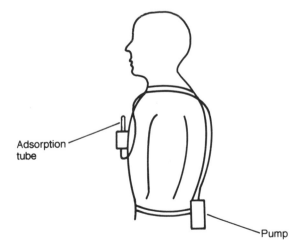

Fig. 6.2e. *Personal sampling*

Desorption of Sample

Transfer of the analyte to the chromatograph is either by thermal desorption or by solvent extraction.

The thermal desorption method uses similar equipment to that used in the analysis of aqueous organic compounds using purge-and-trap techniques (Section 4.2.2).

Solvent extraction requires mixing the adsorbent with a fixed volume of solvent to extract the analyte and injection of the extract into the gas chromatograph.

∏   What problems can you foresee in the use of these sampling methods in quantitative analysis?

The major problem is in the absorption and desorption efficiencies of the sampling. Although, with standard methods, adsorption can be assumed to be 100%, desorption may be less than this and will be different for each compound. Standard analytical methods (e.g. UK MDHS series) recommend specific adsorbents to use for each analyte, but even so, the desorption efficiency has to be measured for each new batch of tubes and has to be included in the analytical calculations.

A second problem is the possibility of overloading the adsorbent. The theoretical capacity of the adsorbent (known as the 'breakthrough volume') can be found either from published tables or by experimental determination, passing a gas of known composition through the tube and monitoring the effluent air with a flame ionisation or similar detector. This volume is, however, influenced by many factors including the presence of water vapour, other organic compounds and temperature, and so will be different for each analysis.

Chromatographic Analysis

The chromatographic separation is usually straightforward using standard columns. There is usually little problem with regard to detector sensitivity for monitoring internal atmospheres and flame ionisation detection is often used. The only difficulty is in the choice of extraction solvent. Solvents which do not show a response with flame ionisation detector (e.g. carbon disulphide, which is toxic and has a low flash point) are often hazardous materials in their own right!

For routine analyses, quantification is by comparison of peak areas with those of standard solutions injected into the chromatograph after correcting for the desorption efficiency.

The desorption efficiency is ideally calculated by analysis of a standard gas mixture, but this is not always practicable. A less rigorous approach is to inject a known quantity of pure compound on to the adsorbent and to measure its recovery on extraction.

Production of Standard Gas Mixtures

Small volumes (up to a few litres) of reference gas can be produced by injecting a known volume of the pure compound, as a liquid, through a septum into an enclosed volume of gas and allowing the liquid to vaporise.

If a continuous flow of reference gas is needed, then dynamic methods are necessary. Permeation tubes are often used. These contain the volatile organic compound within a small PTFE tube which allows slow permeation of the vapour through its walls into a known flow of gas. The rate of diffusion is adjusted by adjustment of the temperature of the tube over the range from ambient to 40 °C. The concentration produced in the gas stream can be calculated from the mass loss of the permeation tube over a given time period and the gas flow rate.

An alternative dynamic technique for gases or volatile liquids is to inject the compound into a gas stream at a constant rate using a syringe pump.

6.2.3. Diffusion Tubes

Most investigations which determine time-weighted average concentrations use one of the two methods described above. There has, however, been a number of applications of the use of diffusion tubes over the past few years, particularly where a large number of sites are being simultaneously monitored. The method incorporates features of the two techniques already described but has the advantage of simple and easy-to-construct samplers which can readily be taken to and left at the sampling site.

The apparatus, as shown in Fig. 6.2f, consists of a short tube (standard dimensions are 7.1 cm length, 0.95 cm inside diameter) which is open at one end and has a liquid adsorbed on stainless-steel mesh at the closed end. The method relies on the natural diffusion of the gas into the liquid.

The reagent is typically exposed to the atmosphere for several weeks, after which the absorbed gas can be determined by standard analytical techniques (e.g. spectrometry). The principle of the technique is that the rate of absorption is determined by the rate of diffusion of the gas along the tube. Fick's law states that the rate of diffusion of a gas is proportional to the concentration gradient. The concentration at the open end of the tube is the ambient concentration. At the closed end of the tube,

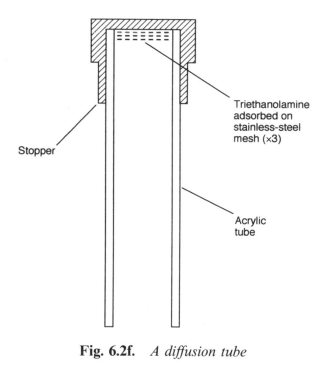

Triethanolamine
adsorbed on
stainless-steel
mesh (×3)

Stopper

Acrylic
tube

Fig. 6.2f.   *A diffusion tube*

it is assumed to be zero as it is being continuously absorbed by the liquid. Hence the rate of diffusion is proportional to the atmospheric concentration.

The original validation of the technique was for internal atmospheres where air currents (which would possibly affect the rate of diffusion and hence the accuracy and precision) would be low. The technique, however, has now been successfully applied to external atmospheres and has been used in major atmospheric investigations in the UK.

The most common use of the technique so far has been for the determination of nitrogen dioxide. The absorbent liquid is triethanolamine, and the analysis is completed by spectrometric analysis (at 550 nm) of nitrate released, using sulphanilamide and N-(1-naphthyl)ethylenediamine hydrochloride.

For a sampling tube of stated dimensions operating at 21 °C:

$$\text{Concentration of NO}_2 = \frac{Q_{\text{NO}_2} \times 1000}{2.3 \times \text{hours exposure}}$$

where Q_{NO_2} = quantity of nitrogen dioxide absorbed (nanomoles). The factor of 2.3 is determined from the known value for the diffusion coefficient of the gas and the tube dimensions.

The precision of the technique is not large (the variance was found to be 10% under ideal conditions). This can be partially compensated for by the low cost of the apparatus, allowing groups of ten or more tubes to be left at each sampling position.

SAQ 6.2a

For routine monitoring of sulphur dioxide in external atmospheres using an absorption train, aqueous hydrogen peroxide is often used as an absorbent;

$$SO_2 + H_2O_2 \rightarrow H_2SO_4$$

rather than the West and Gaeke reagent.

What are the advantages and disadvantages of hydrogen peroxide for large-scale monitoring exercises?

SAQ 6.2a

SAQ 6.2b   Compare and contrast active and passive sampling for monitoring internal atmospheres.

SAQ 6.2c | Alternative methods in the 'Methods of Determination of Hazardous Substances' series (UK) for toluene in atmospheres use solvent extraction and thermal desorption techniques.

What do you see as their relative merits?

6.3. DETERMINATION OF INSTANTANEOUS CONCENTRATIONS

6.3.1. Direct-reading Instruments

Instruments are available to monitor individual gases over the whole range of concentrations we have been discussing. We shall start by looking at instruments designed to be transportable to the monitoring site for measuring ambient concentrations and later discuss methods for workplace and personal monitoring.

Instruments for atmospheric ambient monitoring are often based on spectrometric techniques (chemiluminescence, infrared, fluorescence).

∏ Which of these techniques are potentially the most sensitive and so most suitable for low concentrations found in ambient air?

The techniques involving light emission (chemiluminescence and fluorescence) are potentially the most sensitive. Atmospheric nitrogen oxides, sulphur dioxide and ozone may be analysed by these methods.

Chemiluminescence and Fluorescence

The chemiluminescent method used for nitrogen oxides is based on the reactions

$$NO + O_3 \longrightarrow NO_2^* + O_2$$

$$NO_2^* \longrightarrow NO_2 + h\nu$$

$$\lambda = 600\text{--}875 \text{ nm}$$

Ozone, generated within the instrument, is mixed with the sample under reduced pressure and the light emission is monitored with a photomultiplier, giving a measurement of the nitric oxide concentration of the sample. Total nitrogen oxides can be analysed by thermal conversion of nitrogen dioxide to nitric oxide before analysis. The nitrogen dioxide concentration is then calculated by difference from the two readings. This is the EPA reference method for NO_2 and is also the specified method in EC legislation. Detection limits are approximately 10 ppb $(18 \ \mu g \ m^{-3})$.

The same reaction can also be used to monitor atmospheric ozone. A second chemiluminescent method can also be used, based on the reaction of ozone with ethylene and monitoring light emission at 430 nm. This method has the advantage of little interference from the presence of NO. Detection limits are approximately 1 ppb $(2 \ \mu g \ m^{-3})$.

Sulphur dioxide can be measured without chemical pretreatment by gas-phase fluorescence spectrometry, giving a limit of detection of 2 ppb $(5 \ \mu g \ m^{-3})$.

∏   What method for the production of calibration gases, which has already been discussed, could be incorporated into the instruments.

Calibration methods often use standard gas mixtures generated using permeation tubes (Section 6.2.2).

Infrared Spectrometry

Infrared absorption spectrometry is commonly used in workplaces for monitoring a wide variety of inorganic gases and organic vapours. The spectra can be highly complex and each molecule gives a unique absorption pattern (Fig. 6.3a).

As in other areas of the electromagnetic spectrum, the Beer–Lambert law applies for the absorption of radiation.

If you look back at the mathematical formulation of the law (Section 3.4.1), you will find that the absorbance of radiation is proportional to the molar concentration of the absorbing species and cell path length. In order to maximise the sensitivity for low-concentration gases, long path lengths need to be used. This can be achieved by having large sample cells (up to 1 m) and also reflecting the radiation many times through the cell, giving a total path length of up to 50 m. Instruments relying on absorption spectrometry for gas analysis tend to be bulky, even though they may still be classed as 'portable'.

Instruments may be of similar design to the complex spectrometers with which you will be familiar in the laboratory, which often measure the absorption of radiation after separating the infrared radiation into its spectral components. These are termed dispersive infrared spectrometers. By monitoring the absorption at different wavelengths, a large number of gases can be analysed by a single instrument. One manufacturer pre-programmes the instrument to be able to analyse more than 100 gases and vapours.

An alternative design is sometimes used where no spectral separation is necessary. These are known as non-dispersive spectrometers (Fig. 6.3b).

You may wish to deduce how the instrument operates before reading the following paragraph.

When a molecule in a gas absorbs infrared radiation, the net effect is to heat the gas. It can only absorb radiation at the frequencies which are specific to the molecule. The gas present in the detector cell will heat up and expand only if radiation of a suitable wavelength enters the cell. There is no impediment for suitable radiation on the left-hand side of

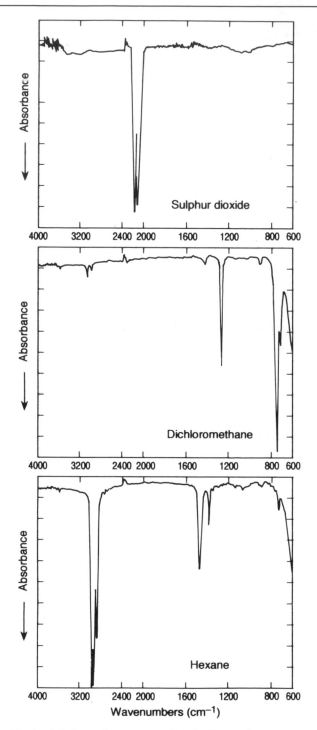

Fig. 6.3a. *Typical infrared spectra of volatile and gaseous compounds*

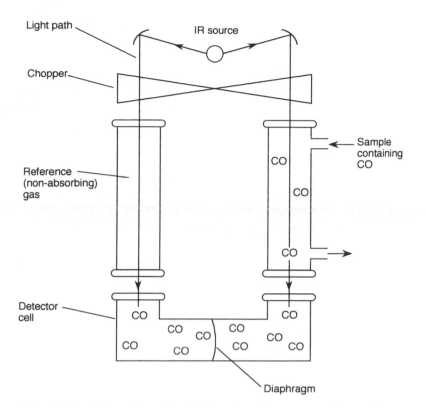

Fig. 6.3b. *Non-dispersive infrared carbon monoxide analyser*

the cell to reach the detector, and the gas will heat up in the detector cell. On the right-hand side some of the radiation at the carbon monoxide specific wavelengths will already have been absorbed by the carbon monoxide in the sample. The heating of the detector cell will be lowered. The net effect is to distort the diaphragm towards the right-hand side. When the beams are turned off the diaphragm will return to its original shape. By chopping the beams an oscillation will be produced. The size of the oscillation will measure the concentration of the carbon monoxide in the gas.

Non-dispersive instruments are available for a number of gases, including carbon monoxide (EPA reference method), carbon dioxide, sulphur dioxide, acetylene, methane and water vapour. Although each is sold as a separate instrument, remember that the only major difference is the gas within the detector cell. It is this gas which gives the instrument its specificity.

∏   Infrared spectrometry is a very suitable method for the analysis of gases when there are few absorbing species, but there may be difficulties when there are many species. Why do you think this is the case? What alternative method do you think would be more suitable?

The complex nature of infrared spectra means that there is the possibility of overlap of absorptions in multi-component mixtures. A chromatographic method (Sections 6.2.2 and 6.3.3), which separates the components before quantification would be more suitable.

Electrochemical Sensors

As we have already found, there is often a need for personal monitoring within a workplace environment, particularly for industrial gases (e.g. CO_2, Cl_2, HCN, HCl, H_2S, SO_2) which may quickly build up in atmospheres through leakages, or build up in unventilated areas. Personal monitors are available for individual gases based on electrochemical sensing, a different sensing head being required for each gas. The reaction of the analyte gas at an electrode produces a current which is proportional to its gas-phase concentration. The monitors normally possess a concentration display and also an audible alarm if the preset maximum concentration is exceeded.

∏   Portable site instruments use infrared absorption techniques whereas personal instruments use solid-state electrochemical methods. Can you explain why different methods are used for each application?

Infrared absorption varies with concentration and path length according to the Beer–Lambert law. For maximum sensitivity for low-concentration gases, long path lengths (several metres) are necessary. This leads to bulky instrumentation, even when multiple reflections are achieved in the sample cell. In addition, optical components tend to be heavy, and need careful alignment, very undesirable features for personal instruments.

On the other hand, solid-state electronics are lightweight and rugged, making these ideal for personal monitors. Electrochemical techniques, however, do require more frequent calibration than spectrometric methods, making them less appropriate for long-term background analysis.

6.3.2. Gas Detector Tubes

Hand-held and easy-to-use instruments are often needed for monitoring internal atmospheres where high concentrations of hazardous gas can quickly accumulate. One simple type of instrument uses gas detector tubes.

Gas detector tubes are available from a number of manufacturers for most common inorganic gases and volatile organic compounds. Trade names include Draeger, Gastec and Kitagawa. They are constructed of glass, are several centimetres in length and are packed with an analyte-specific reagent adsorbed on an inert solid.

A fixed volume of gas is drawn through the reaction tube using a hand pump. This may be a bellows-type or piston pump according to the manufacturer. A typical example is shown in Fig. 6.3c.

The sampling time is a few seconds, during which a colour develops from the sampling end of the tube. At the end of the sampling period the colour should extend along a fraction of the length of the tube. The

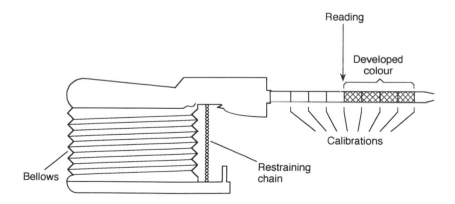

Fig. 6.3c. *Gas sampling tube with bellows*

tubes are precalibrated with a concentration scale on the glass surface so that the distance the colour has travelled can be directly related to the gas concentration.

The colour may be produced by a number of methods. In some cases a coloured product is formed from colourless reagents. Hydrogen sulphide detection tubes contain a colourless lead salt adsorbed on silica gel. The product is black lead sulphide.

$$Pb^{2+} + H_2S \longrightarrow PbS + 2H^+$$

In other cases the colour is due to an indicator change. Detector tubes for carbon dioxide contain hydrazine as reactant and crystal violet as redox indicator. The reaction with carbon dioxide causes the indicator to change colour to purple.

$$CO_2 + N_2H_4 \longrightarrow H_2N-NH-CO_2H$$

A range of tubes are available for each gas to accommodate different concentration ranges. The available ranges are typical of internal atmospheres and emission concentrations (ppm to percentage levels), but can in some cases be extended to lower levels. This is achieved by increasing the gas volume sampled using a continuous pump rather than bellows. Under these circumstances care has to be taken that the reagent will not be stripped from the support or oxidised (check in Section 6.2.1 for similar potential problems encountered with reagents in absorption trains). The tubes are not generally used for sampling periods of longer than a few minutes, and hence are not suitable for very low level contamination.

Two further problems need to be considered before the application of these tubes:

1. Precision

The relative standard deviation varies from compound to compound. In the most favourable cases (e.g. hydrogen sulphide detection), where the chemical reaction proceeds rapidly, the relative standard deviation is 5–10%. In less favourable cases (e.g. mercury detection), a relative standard deviation of 20–30% may be found.

2. Interferences

These are well known for inorganic gas detection and specified in manufacturers' literature. Sometimes a separate zone of reactive solid is included in the detector tube to remove potential common interferences before they reach the calibrated layer. Carbon monoxide tubes contain a zone of chromium(VI) to remove hydrogen sulphide, benzene and other organics. Interferences may, however, be more serious for detection of specific organic compounds. As an example, adjacent members of the homologous series give positive indications on hexane tubes.

6.3.3. Gas Chromatography

We are considering here GC methods where a sample of gas is introduced directly into the chromatograph without preconcentration. This method finds application for the analysis of inorganic gases (e.g. O_2, N_2, CO, CO_2), particularly in exhausts or flues, and also for atmospheres containing mixtures of volatile organic compounds.

The chromatograph may be situated in a laboratory, but could also be a portable design which could be carried to the monitoring site, or it could be permanently positioned at the sampling point, away from the laboratory.

The following discussion is centred on the use of a laboratory based instrument, with additional comments if there are any differences in portable or site-based instruments.

∏   In what ways do you think a portable instrument may be different from a laboratory instrument?

Changes would have to include decreases in size, mass and number of gases and utilities used to make the chromatograph portable. Most manufacturers achieve this by use of a chromatographic column which can separate the components at, or near, ambient temperature. This removes the need for a high-temperature oven. Detectors which do not need additional gas supplies are also favoured, although flame ionisation is sometimes used for organic analyses.

Sampling

A large variety of containers may be used for sampling the gas (Fig. 6.3d). Portable instruments usually include a small pump to draw in gas through a sampling tube.

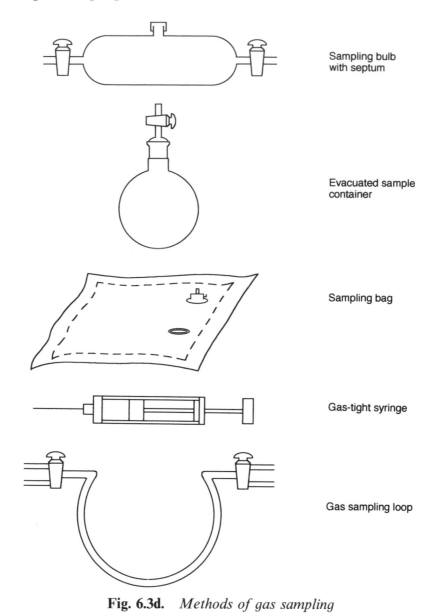

Sampling bulb with septum

Evacuated sample container

Sampling bag

Gas-tight syringe

Gas sampling loop

Fig. 6.3d.   *Methods of gas sampling*

∏   What problems do you see in sampling gases and injecting them into a gas chromatograph?

Large sample vessels are necessary, typically of several hundred millilitres.

It is difficult to check for leakage/contamination of the sample.

Minor components may be lost by reaction on the walls of the vessel.

Injection of large volumes of gas into the chromatograph disturbs the carrier gas flow.

Chromatographic Analysis

Gas–solid chromatography is used for the separation of inorganic gases and low molecular mass organic compounds. Several types of stationary phase are used.

Molecular sieves are often used for permanent gases. These separate gases in order of their molecular size (Fig. 6.3e).

Unfortunately, one of the most important components of flue gas, carbon dioxide, is permanently adsorbed by the molecular sieve. A silica

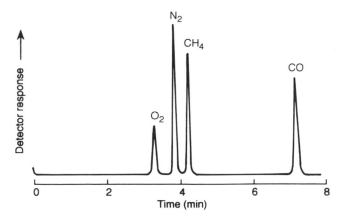

Fig. 6.3e.   *Separation of a gas mixture on a 5A molecular sieve column*

gel column, which separates by adsorption, is needed for this gas. Note that the other common inorganic gases are not well separated on this column so a complete flue gas analysis would require both columns.

Organic porous polymer adsorbents may be used for both low molecular mass organic compounds and inorganic gases. Several manufacturers produce a series of stationary phases, e.g. Porapak series. The chromatographic separation of the gases can be optimised by suitable choice of stationary phase within the series.

Conventional gas–liquid chromatographic columns may be used for volatile organic compound separations.

∏   What gas chromatographic detector have you come across which is suitable for inorganic gases such as oxygen, nitrogen and carbon dioxide?

The thermal conductivity detector (katharometer) is suitable for all gases.

Flame ionisation detection, which is often said to respond universally, will not easily detect most inorganic gases.

The thermal conductivity detector has relatively low sensitivity and so cannot be used for trace analysis, the lower limits of detection being in the region of a few hundred parts per million. The greatest sensitivity can be achieved by use of a low molecular mass carrier gas. Hydrogen would give the greatest sensitivity but its use is often discouraged on safety grounds. Helium as carrier gas gives a slightly lower sensitivity than hydrogen. However, the cost of helium varies enormously worldwide and in some countries is a very expensive option.

An alternative procedure is possible for carbon monoxide detection where the gas is reduced to methane using a nickel catalyst. This may then be detected with high sensitivity using flame ionisation detection.

Flame ionisation detection can also be used for monitoring organic components, but has the disadvantage for portable and site instruments in requiring the maintenance of the detector flame. Thermal conductivity detection can be used as an alternative for high-concentration components. At least one manufacturer uses photoionisation detection for trace organic analysis.

∏   Can you think of a method for determining the total organic vapour content of an atmosphere without determining each component separately?

One method would be to inject a sample directly into a flame ionisation detector without passing it through a chromatographic column. The response would be proportional to the total organic content.

Total organic vapour monitors are commercially available which are based on this method. Some portable chromatographs with flame ionisation detection have an option to by-pass the column in order to measure total organic vapour concentrations.

6.3.4. Remote Sensing

We have seen that many instruments are available for gases which can sample and analyse at locations remote from a laboratory. Spectrometric methods are often used. You may wonder why, if you are simply measuring the light absorption of a gas, you need to take a sample at all. Why not simply measure the light absorption through a section of the atmosphere? This is the principle of remote sensing.

Most compounds of environmental concern are found at ppb v/v concentrations. Confirm this by looking back at Fig. 6.1a. You may consider these concentrations too low for analysis by direct absorption measurements, but you should remember that extremely long path lengths can be used to compensate for the low concentrations.

First we must choose a wave of light which is absorbed by the analyte and not by other components in the atmosphere. A few atmosphere components show characteristic absorptions in the ultraviolet region of the spectrum. These include:

sulphur dioxide;

nitrogen dioxide;

ozone.

The infrared region of the spectrum can be used for monitoring localised pollution of compounds which do not absorb in the ultraviolet region. More widespread use of infrared spectrometry is hindered by strong absorption of the atmosphere (by carbon dioxide and water) over much of the infrared range.

The instrumentation needed for this seemingly simple technique is complex. The light source is a pulsed laser. Figure 6.5f shows a configuration where the light detector is positioned close to the laser. This method is known as LIDAR (light detection and ranging).

The light reaches the detector over a slightly more extended time period than the original pulse length owing to the different path lengths in the atmosphere. Looking at Fig. 6.3f, the light travelling on path ABD will reach the detector before the light on path ACD. The intensity of light reaching the detector is measured over the complete return time period. This information can be processed to give the concentration of the absorbing species over each of the light paths, and this can be built up to give a concentration profile over the complete sampling range.

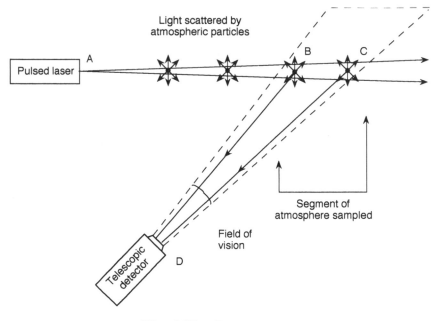

Fig. 6.3f.   *Remote sensing*

SAQ 6.3a   Compare the potential uses of absorption trains, adsorption on solids and direct-reading instruments as methods of analysis of atmospheric samples.

SAQ 6.3b   Which techniques would you use for the following analyses?:
(*a*) nitrogen dioxide in the external atmosphere at several locations;
(*b*) an organic solvent in a laboratory atmosphere;
(*c*) carbon monoxide, to protect a worker in an area where there may be rapid increases in concentration.

SAQ 6.3b

Summary

Concern over gaseous pollutants includes not only those found in external (outdoor) atmospheres, but also internal (indoor) atmospheres, since both can have an effect on human health. The types of pollutant found in these two areas can differ in chemical type and in concentration, higher concentrations often being found with internal atmospheres.

The concentrations can often change rapidly with time. Concentrations averaged over a fixed time period (time-weighted averages) are the most appropriate measurement for long-term investigations. A detailed study of a pollution incident would, however, require instantaneous concentration measurements. Methods have been described for both of these types of measurement, and include techniques which determine concentrations directly in the field and techniques requiring the analysis to be completed in the laboratory.

Objectives

You should now be able to:

- list major components in external atmospheres and appreciate their importance in the environment;

- understand the difference in type and concentration of pollutants in external and internal environments and the difference in approach needed for their analysis;

- understand the nocooolty of personal sampling;

- describe and compare the analytical methods available for external and internal atmospheres:

- appreciate the availability of portable instruments for gas analysis and the possibilities of remote sensing.

SAQs AND RESPONSES FOR PART SIX

SAQ 6.1a   What is the difference in meaning of the term 'parts per million' when applied to gas concentrations and aqueous concentrations?

Response

The definition of the term for gas concentrations is very precise, referring to measurements made as volumes. An atmosphere containing 20 ppm of sulphur dioxide would contain 20 μl of gas per litre of atmosphere. A complete statement of the unit should be parts per million *volume/volume*.

When the term is applied to aqueous concentrations it is often used interchangeably with mg l^{-1}. This would give the complete statement of the unit as parts per million mass/volume! Since 1 litre of water containing little dissolved material has a mass of 1000 g, this would become

1 ppm = 1 mg analyte per 1000 g water

i.e. parts per million mass/mass for water samples.

SAQ 6.1b	Briefly summarise the expected concentration ranges of pollutants in external and internal atmospheres and in exhaust gases. Suggest reasons why we may find there are sometimes different analytical methods used for external and internal atmospheres.

Response

	Typical pollutant concentrations (v/v)
External atmospheres	parts per billion (ppb)–parts per million (ppm)
Internal atmospheres	parts per million
Exhausts or flue gases	parts per million–parts per hundred (%)

1. Different Concentration Ranges

The concentrations in the two atmospheres span a range of 10^7. It is not surprising that some methods are more readily applicable to the low concentrations and others to higher concentrations.

2. Different Analytes

Unless you are concerned with highly localised pollution, the number of gaseous pollutants which can build up to detectable levels in the external atmosphere is small. Although Fig. 6.1a is by no means a comprehensive list, it does give an indication of the type of compounds which may be present—simple inorganic gases, a few stable organic compounds and a number of photochemically generated species. A greater diversity can build up in internal atmospheres, and in particular many organic compounds. These will require different monitoring techniques.

3. Concern over Human Health

You might expect since internal atmosphere monitoring is largely concerned with human health, that instantaneous concentration measurements (or short-term time-averaged concentrations) will be important, alongside longer term time-averaged values. Longer term averaged values often predominate in importance for external atmospheres. Different methods may be needed for the two types of determination.

4. Sampling Difficulties

We have not discussed this point before, but you should realise that air currents are usually weaker and more stable indoors than in the external environment. Representative samples may be easier to obtain in an internal environment. We shall find that both the accuracy and the precision of at least one of the methods we shall be discussing are lowered by strong air currents. When a new analytical technique is introduced there is sometimes a progression of validation first for internal atmospheres and subsequently for external atmospheres.

SAQ 6.2a

For routine monitoring of sulphur dioxide in external atmospheres using an absorption train, aqueous hydrogen peroxide is often used as an absorbent;

$$SO_2 + H_2O_2 \rightarrow H_2SO_4$$

rather than the West and Gaeke reagent.

What are the advantages and disadvantages of hydrogen peroxide for large-scale monitoring exercises?

Response

1. Hydrogen peroxide solution is readily available.

2. The reaction product may be estimated by volumetric titration (e.g. using standard sodium hydroxide solution), removing the necessity for spectrometers and well equipped laboratories.

3. The method is, however, non-specific. Any atmospheric component which will dissolve to form a strong acid or can be oxidised to a strong acid will be included in the final analysis.

For ambient air monitoring, potential interferences are likely to be at lower concentration than the sulphur dioxide but for other analyses (e.g. flue gases) this may not necessarily be the case.

SAQ 6.2b | Compare and contrast active and passive sampling for monitoring internal atmospheres.

Response

Passive sampling techniques will require longer sampling times than the corresponding active sampling techniques, since they rely on gas diffusion. The minimum sampling time is several hours even for internal atmospheres, and so would be of little use in short-term monitoring.

Active sampling techniques have greater flexibility. The sampling rate can be adjusted, within limits, according to the application, making both long- and short-term monitoring possible. However, if used for personal monitoring, the pumps necessary for active sampling can be inconvenient for the wearer, and most would prefer to be monitored by passive sampling techniques.

SAQ 6.2c

> Alternative methods in the 'Methods of Determination of Hazardous Substances' series (UK) for toluene in atmospheres use solvent extraction and thermal desorption techniques.
>
> What do you see as their relative merits?

Response

Solvent extraction can use standard laboratory apparatus. It is, however, time consuming and can use potentially hazardous solvents.

Thermal desorption methods need specialised instruments, but these minimise laboratory manipulation. The sensitivity can be higher than for solvent extraction since the whole sample is introduced into the chromatograph in a single desorption. Only a small fraction of the extract is injected into the chromatograph in the solvent extraction method. However, replicate determinations are not possible with thermal desorption. This is easily achievable using solvent extraction.

SAQ 6.3a

> Compare the potential uses of absorption trains, adsorption on solids and direct-reading instruments as methods of analysis of atmospheric samples.

Response

1. Absorption trains can sample continuously over a 24 h or 8 h period to obtain a time-average value from a single analysis, are less expensive than single instruments, leading to the possibility of simultaneous sampling at different locations, and can be used as reference methods for other techniques.

2. Adsorption on solids with subsequent analysis by gas chromatography is most frequently used when there is a central analytical facility for personal and multiple site monitoring.

3. Direct-reading instruments are used for continuous monitoring of atmospheres, at a limited number of sites. Their high expense would limit their use to more comprehensive monitoring exercises.

SAQ 6.3b | Which techniques would you use for the following analyses?:
(a) nitrogen dioxide in the external atmosphere at several locations;
(b) an organic solvent in a laboratory atmosphere;
(c) carbon monoxide, to protect a worker in an area where there may be rapid increases in concentration.

Response

(a) An absorption train could be used for the analyses. If a large number of sites were involved, passive samples could be used as an alternative, but with lower precision. A specific NO_2 analyser (e.g. chemiluminescence) would be more appropriate for continuous single-site analysis.

(b) The most reliable method would be sampling the atmosphere using adsorption tubes with subsequent gas chromatographic analysis. Gas detector tubes are available for common solvents, which would give almost instantaneous determinations, but care would have to be taken over possible interferences from other solvents which may be present in the laboratory.

(c) A personal monitor responding to carbon monoxide would provide protection. Continuous monitors (e.g. an infrared spectrometer) could also be located in the most hazardous areas.

7. Atmospheric Analysis— Particulates

OVERVIEW

This part of the book introduces you to the sampling and analysis of atmospheric particulate material. Analytical methods are divided into those requiring sample dissolution prior to analysis and those which can analyse solid material directly without a dissolution stage. The solid-phase analytical techniques are briefly described.

7.1. INTRODUCTION

Have we now discussed analysis of all the major components of the atmosphere? Certainly not! We have only looked at gases and vapours. An equally important component is the particulate matter. Particulates are natural component of the atmosphere. They include:

— Condensation products from natural combustion (forest fires, volcanoes).

— Products of reaction of trace gases (ammonium chloride, sulphate and nitrate salts).

— Material dispersed from earth's surface (salt spray from oceans, mineral dust from continental land mass).

In addition to these is the particulate material introduced by man. This can predominate in urban atmospheres, the major sources being combustion and incineration processes.

Particulates have an important role in the chemistry of the atmosphere. Many atmospheric reactions which, at first sight, appear to take place in the gas phase, occur either on the surface of the particulate matter or in the liquid phase in water adsorbed on the surface of the particle.

Let us take as an example the smogs which regularly affected London until the mid-1950s. The primary components of the smog were sulphur dioxide and particulate matter, both derived from coal combustion. In one major incident in 1952 approximately 3000 deaths in one week were attributable to the smog. The maximum concentration of sulphur dioxide was found to be 3.8 mg m^{-3} (1.34 ppm). This concentration, which was very much higher than normal, has been shown to cause no adverse effects in man without the presence of particulate matter. A strong synergistic effect was indicated. The particulate matter provided a surface for the liquid-phase oxidation of sulphur dioxide to sulphuric acid, which remained adsorbed on the surface of the particle. The size distribution of the particulate matter was such that, on inhalation, a fraction of the particles lodged in the lungs. The irritation caused by the particulate material was increased by the adsorbed layer of sulphuric acid.

Atmospheric transport in the form of particulates is one of the major methods for the dispersal of pollutants. We have seen in Part 2 that a significant route for dispersal of lead is via the atmosphere (the lead being transported predominantly as inorganic salts). Similar routes can be constructed for the other metals of environmental concern.

Low-volatility organic material occurs in the atmosphere partly in the vapour state and partly in the solid phase, either as organic particulates or adsorbed on inorganic particulates. Any consideration of the transport of organics needs to take both vaporised and particulate fractions into account.

Let us consider what measurements may be useful for the characterisation of the particulate content of an atmospheric sample.

   (i) A preliminary measurement would be the total particulate concentration. This is a measurement of the mass of solid extracted from a fixed volume of atmosphere by filtration or by other methods

Typical values are:

70 μg m^{-3} rural air

300 μg m^{-3} urban air

1 mg m^{-3} factory workshop

100 mg m^{-3} power station flue gases

(ii) The second consideration is the analytical composition. Often this is simply elemental analysis, particularly for metals. The analytical task can be more difficult than we found for aquatic samples since the inorganic component of the particulate material may be highly insoluble, particularly if present as silicate salts. All the analytical techniques we have so far come across involve sample dissolution. Two approaches are possible for 'insoluble' particulates. Extreme conditions may be used to dissolve the sample, followed by the methods for metals which we have already discussed. The alternative approach is the use of techniques which do not require sample dissolution.

(iii) The particle size distribution is often also determined.

∏   Why do you think particle size is important?

1. *Transport.* The residence time of a particle in the atmosphere is dependent on its size. The greater the size, the more rapidly deposition from the atmosphere occurs (Fig. 7.1a). Particles less than 0.1 μm in diameter can be considered to be capable of permanent suspension.

2. *Differences in physiological properties.* The smaller the particle size the greater is the possibility of the particle entering the gas exchange region of the lungs. It is this material which will have the greatest potential physiological effect. This fraction of the particulate matter is termed 'respirable' dust, and as a general guide, would refer to material below approximately 5 μm. The corresponding term for the total inhalable fraction is 'inspirable' dust.

3. *Distribution of chemical species.* If you are studying emissions from a particular industrial process you may find that the particulate

Particulate diameter (μm)	Commonly used term	Sedimentation velocity in still air (cm s^{-1})
< 0.1	Fume	Negligible
0.1		8×10^{-5}
1	Smoke	4×10^{-3}
10		0.3
10		0.3
100	Dust	25
> 100	Grit	> 25

Fig. 7.1a. *Classification of particulate material*

matter is often within a narrow size range. Fractionation of the dust sample may then constitute an essential part of the analytical procedure.

Due consideration has to be taken in all the following methods of the low particulate concentrations found in the atmosphere. Even with long sampling times in heavily polluted atmospheres you will only be dealing with milligram quantities of sample, making a very exacting analytical task. As in other sections, we shall discuss sampling methods first, then the analytical methods.

7.2.   SAMPLING METHODS

The importance of careful sampling strategies has been stressed throughout the book, but perhaps nowhere is it more important than with particulate sampling. Concentrations can vary rapidly with time and location. In internal atmospheres there is often a measurable vertical variation even over a few centimetres. This leads to an emphasis towards personal sampling to assess the exposure of an individual rather than comprehensive surveys of background levels. For external atmospheres, however, background concentrations using large-throughput (high-volume) samplers remains the most appropriate method.

7.2.1. High-volume Samplers

The air sample is drawn through a large-diameter membrane filter (20–25 cm) typically at $75 \text{ m}^3 \text{ h}^{-1}$. The construction of the sampler is easiest to understand when you discover that the earliest types were modified from commercial vacuum cleaners, i.e. it is simply a fan behind a filter holder. Nowadays, of course, purpose-built apparatus is readily available. Typical sampling times range from 1 h for contaminated urban atmospheres to 12 h for clean rural atmospheres, with shorter times possible for internal atmospheres.

The choice of filter is based on,

1. retention of correct particle size range;

2. absence of trace impurities in the filter;

3. compatibility with the subsequent analytical procedure. Some procedures require the total combustion of the filter, others its dissolution.

Cellulose filters should be used for metals and inorganic anions and glass-fibre filters (or under some circumstances silica filters) for organics.

7.2.2. Personal Samplers

A filter holder is clipped to the lapel and the pump around the waist. The pump is similar in design to those used for organic gas sampling as described in Part 6, with one difference—dust sampling is at the higher rate of approximately 2 l min^{-1} through a 25 mm filter. Filters are glass-fibre if simply a total particulate mass is required. Other filter material may be used for elemental analyses, depending on the subsequent analytical procedure.

This apparatus will produce a representative sample of total inhalable (inspirable) dust. If a sample of respirable dust is required, then a preselector is necessary to ensure that only particulates of the correct size range reach the filter. A cyclone elutriator (Fig. 7.2a) may be used for this purpose.

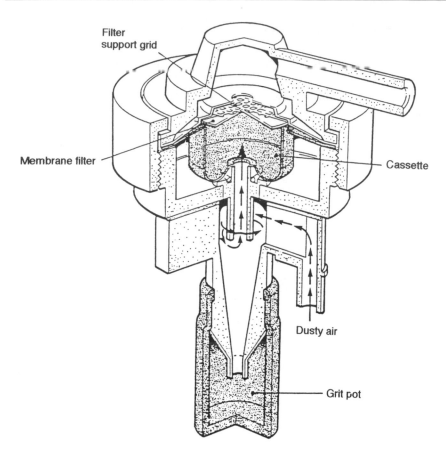

Fig. 7.2a.   *A cyclone elutriator*

The gas is spiralled through a conical container in such away that particulates outside the required size range fall into a container at the base of the elutriator, rather than passing on to the filter.

7.2.3.   Cascade Impactors

The previous two methods used filtration for collection of the particulate material. Cascade impactors rely on adhesion of particulates on a surface. The particulates are fractionated according to their mass. A typical apparatus is shown in Fig. 7.2b.

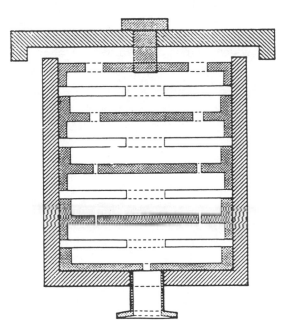

Fig. 7.2b. *A cascade impactor*

Air is drawn through the device at a constant rate to impact on a number of targets coated with petroleum or glycerine jelly. By constriction of the flow before each target the linear velocity of the air increases. Particles adhere to the targets if they impact above a specific momentum (momentum = mass × velocity). Since the air velocity increases through the apparatus successively smaller particles will adhere to each successive surface. A typical operating range is 0.5–200 μm.

Typical operating flow rates would be 1 m³ h⁻¹, producing only a few micrograms of sample in each fraction per hour of operation when sampling a typical urban atmosphere.

7.2.4. Further Considerations for Organic Compounds

If sampling is to represent the total organic content of the atmosphere, it has to accommodate both solid and vapour phases. After passing through a filter to remove particulates, the gas is drawn through an adsorbent (Section 6.2) to extract the vapour-phase component. The analysis of the two phases can then proceed separately.

SAQ 7.2a

> Consider a studio glass-making furnace in a small room for which there is concern over a technician's exposure to the particulate lead emissions. How would you set about sampling the atmosphere?

7.3. ANALYTICAL METHODS INVOLVING SAMPLE DISSOLUTION

7.3.1. Metals

A first step in any analytical procedure should be to consider the probable composition of the sample. It is a vital step for particulate analysis which will allow the correct choice of dissolution technique. If the composition of the sample is unknown, as would be the case for many external atmosphere samples, hydrofluoric acid, which is capable of dissolving silicates, may be required. This acid causes severe burns and attacks glass apparatus (the silica structure of the glass is closely related to the insoluble silicates which you may be trying to dissolve). Teflon apparatus is required and the analyses should be performed in a hydrogen fluoride-resistant fume cupboard. You may now be able to see why this method is avoided whenever possible.

If the composition of the dust sample is known (as may be the case with samples from workplace environments) the dissolution may be less severe, according to the known solubility of the sample. Dilute acid, mild oxidising agents or even water may be all that is necessary for dissolution.

To illustrate the difference, let us look at two standard methods for the analysis of lead in dust.

1. The SCOPE procedure of the International Council of Scientific Unions involves the following stages:

   Collection of the particulate matter in a glass-fibre filter (twice washed with distilled water).

   Warming with hydrofluoric acid until the liquid is almost evaporated.

   Repeating with nitric acid.

   Diluting to volume with distilled water.

   This procedure solubilises the filter in addition to the sample.

2. The UK Methods for the Determination of Hazardous Substances (MDHS) procedure for internal atmospheres assumes that lead is in a more easily soluble form and uses a simpler one-stage procedure of warming with nitric acid/hydrogen peroxide. The filter-paper remains undissolved.

Once the sample has been dissolved, the analysis can proceed by a number of methods available for metal ions in solution.

∏   Which two methods have you come across which may be most suitable for the routine analysis of metals in particulates?

Atomic absorption spectrometry.

Visible/ultraviolet absorption spectrometry.

For less routine analysis, and particularly for the analysis of metals at low concentrations, other techniques may sometimes be used. These include inductively coupled plasma optical emission, inductively coupled plasma mass spectrometric, flame atomic emission and atomic fluorescence techniques.

The sensitivity of each technique is different for each element. Comparative data are shown in Fig. 7.3a. Take care when using such a table as

Element	Furnace AAS[a]	ICP-OES	ICP-MS	Flame atomic emission	Atomic fluorescence
Ca	0.03	0.07	2	0.1	0.001
Cd	0.01	1	0.003	800	0.01
Mn	0.05	0.7	0.002	5	2
Pb	0.3	8	0.001	100	10

[a]20 μl samples are assumed for furnace AAS.

Fig. 7.3a. *Comparative detection limits ($\mu g \, l^{-1}$) of atomic spectrometric techniques*

limits may change between equipment manufacturers and with improvements in instrumentation.

If you assume a 1 m^3 air sample and the metal extracted into 5 ml of acid, the table covers a range of detection limits from 4 μg m^{-3} (Cd using atomic emission) to 5×10^{-6} μg m^{-3} (Ca using atomic fluorescence and Pb using ICP-MS).

7.3.2.   Organic Compounds

Simple determination of organic content may be by analysis of total organic carbon (Section 3.3.3) or by mass loss after extraction with an organic solvent. The components of the extract can then be determined by the spectrometric and chromatographic methods described in Part 4.

SAQ 7.3a

Atomic absorption and ultraviolet/visible spectrometry are often specified as alternatives in standard methods for the analysis of metals as particulates in workplace atmospheres. This contrasts with the predominance of atomic absorption for analysis of aqueous samples.

What are possible reasons for this difference?

SAQ 7.3a

SAQ 7.3b What feature of the sampling and analysis of atmospheric particulates could account for the variety of techniques used for analysis of low concentrations of metals?

7.4. DIRECT ANALYSIS OF SOLIDS

We shall briefly discuss a number of representative techniques.

The first three are methods for elemental analysis using equipment which will only be available in specialist laboratories. An example of a solid-state method using a readily available laboratory instrument is then discussed. The final section briefly mentions the specialised techniques used for asbestos analysis.

7.4.1. X-ray Fluorescence

This technique is based on the irradiation of an atom with x-rays leading to the ejection of an electron from an inner shell. Outer shell electrons cascade to the inner shell to fill the vacancy, emitting x-rays. The wavelength of this radiation is related to the atomic number of the nucleus according to the equation

$$1/\lambda = kz$$

where

λ = wavelength of radiation;
k = constant;
z = atomic number;

i.e. elements emit radiation at characteristic wavelengths. Absorption and emission occur predominantly in the first few surface layers of atoms. With suitable corrections for matrix effects, which may include the preparation of standards with composition as close as possible to the sample, the intensity is proportional to the concentration of the element.

Two types of instrument are available, which differ according to how the fluorescent radiation is analysed.

Wavelength-dispersive instruments measure the emission at each wavelength sequentially, using diffraction from a rotating crystal to direct individual wavelengths to the detector.

Energy-dispersive instruments measure the whole of the fluorescence simultaneously at the detector. The contributions from each wavelength are separated electronically. This type of instrument is more convenient to use and produces more rapid analyses, but has a slightly lower sensitivity.

A typical spectrum is shown in Fig. 7.4a.

Elements above atomic number 40 can be routinely analysed and using vacuum techniques elements from F to Ca can also be measured. Particulate samples collected on filter-paper can be analysed without pretreatment. Detection limits for elements vary widely, but for airborne

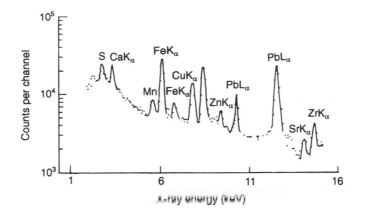

Fig. 7.4a. *X-ray fluorescence spectrum of a dust sample*

particles they are of the order of 10^{-2} μg m^{-3}, when expressed as the original atmospheric concentration.

7.4.2. X-ray Emission

X-rays may also be generated by the bombardment of a sample with fast electrons. The bombardment again causes excitation of inner shell electrons with the subsequent decay back to the ground state, causing x-ray emission. This technique is used in the electron microprobe analyser. The electron beam can be focused on a small area, which can be as small as an individual dust particle. This then gives an extremely powerful technique for assessing a composite dust sample. It is also one of the few techniques capable of quantitative analysis of the low-mass samples produced by a cascade impactor. The instrument may also be used as a conventional microscope. Qualitative analyses of the particles can be made from images produced emitted by x-rays at wavelengths corresponding to individual elements.

7.4.3. Neutron Activation Analysis

The sample is irradiated with neutrons to produce radionuclides of the elements of interest. As the radioactive nucleus decays it emits gamma rays. The intensity of the gamma-ray spectrum can be related to the original concentration in the sample, e.g.

$$_{25}^{55}\text{Mn} + {}_0^1 n \rightarrow {}_{25}^{56}\text{Mn} \rightarrow {}_{26}^{56}\text{Fe} + \beta^- + \gamma \qquad t_{1/2} = 2.58 \text{ h}$$

The technique is highly sensitive, needing as little as 0.1 μg of sample. Detection limits for elements in airborne particulates can be as low as $2 \times 10^{-5} \mu$g m^{-3}. No chemical pretreatment is necessary and the only physical treatment is grinding and homogenisation of large samples. The one major disadvantage is that you need a source of neutrons, usually a nuclear reactor!

7.4.4. Infrared Spectrometry

This method can be used for compounds which have characteristic absorption frequencies well away from those of likely interfering components. Quartz may be determined by this method using absorptions at 780 and 800 cm^{-1}. The sample is introduced into the beam either directly on the filter-paper or after making a pressed disc by grinding the sample with potassium bromide and compressing under high pressure. The absorptions are compared with standards produced from atmospheres containing known quantities of quartz of similar particle size to that of the sample.

7.4.5. Methods for Asbestos Analysis

Asbestos is a term used for any one of a group of fibrous silicate minerals. They possess good heat and electrical insulation properties and have found widespread use in industry. It has, however, become a major environmental hazard. Airborne fibres are capable of being trapped in the lungs. The respiratory disease asbestosis can result, and also a number of forms of cancer.

The detrimental effects on human health are related to the shape and size of individual fibres. Microscopic analysis is essential. The method using optical microscopy involves collection of the particulate material from the atmosphere by filtration, preparation of a microscope slide and then identifying and counting fibres in the microscope field of view. The field of view is then altered and the counting repeated a number of times. Results are expressed in terms of the number of fibres per millilitre of air.

Electron microscopic techniques may also be used, giving the possibility of chemical analysis of individual fibres (Section 7.4.2) as an additional means of identification.

∏   Solid-state analytical techniques appear to offer many advantages over methods requiring sample dissolution. What are the disadvantages which sometimes restrict their use?

1. The techniques avoid the dissolution stage of other procedures. For particulate analysis this can be a difficult process. Sample preparation is, however, still required for some solid-state techniques and also in the production of calibration samples. For small numbers of samples the time saving is then not as great as would first appear.

2. The direct analysis of solid material also poses problems for the analysis of large samples. The material analysed (a few milligrams at most) has to be representative of the whole. We have seen, however, that this can be put to advantage with electron microprobes which are able to analyse individual particles.

3. Some methods, including x-ray fluorescence, only respond to the first few layers of atoms within a sample. Surface layers may have a different composition to the bulk and, without due care, misleading results may be produced. Another problem with x-ray methods is the possibility of matrix effects.

4. Many of the techniques require highly specialised spectrometers which may not be routinely found in general analytical laboratories.

SAQ 7.4a   What criteria would you use to choose the analytical technique for several metal ions in a particulate sample?

SAQ 7.4a

Summary

Particulate material is an essential and natural component of the atmosphere. Much airborne pollution is, however, also in the form of particulate material. Analysis of the material starts with sampling from the atmosphere. This is often by filtration. The method used for the chemical analysis depends on the ease of solubility of the material. If the substance is readily dissolved then the analysis can proceed using techniques already discussed for species in solution. If the substance is more difficult to dissolve then techniques which do not require sample dissolution (solid-state analytical techniques) may be used. Examples of both types of method have been discussed.

Objectives

You should now be able to:

- understand the importance of particulate material in the atmosphere;

- determine appropriate methods of sampling particulates;

- assess the relative merits of analyses involving sample dissolution and methods not requiring a dissolution stage;

- appreciate the range of solid-state analytical techniques which may be available in specialist laboratories.

SAQs AND RESPONSES FOR PART SEVEN

SAQ 7.2a   Consider a studio glass-making furnace in a small room
             for which there is concern over a technician's exposure to
             the particulate lead emissions. How would you set about
             sampling the atmosphere?

Response

The most relevant sampling would be using a personal sampler with the
filter holder attached to the technician's lapel. However, this should be
backed up with static sampling at a number of locations within the
room. The location of the static sampling should be in the area where
the technician is liable to be working and predominantly in areas where
you consider high concentration of particulates to be likely. The areas
of high concentrations will be determined by the air flows in the room
which will be produced by the convection currents from the furnace and
doors, windows and any extraction system. The vertical location should
reflect, if possible, the height of the breathing zone of the technician in
his most usual stance, whether seated or standing.

Sampling should be over as long a period as possible to reflect the
exposure over an 8 h working day. Since large variations in exposure
are possible, monitoring should be repeated for several days.

SAQ 7.3a   Atomic absorption and ultraviolet/visible spectrometry
             are often specified as alternatives in standard methods for
             the analysis of metals as particulates in workplace
             atmospheres. This contrasts with the predominance of
             atomic absorption for analysis of aqueous samples.

             What are possible reasons for this difference?

Response

From a practical point of view, routine analysis will often be performed
in small laboratories close to the workplace being monitored and with

limited facilities. Under such circumstances, ultravisible/violet spectro-metry may be a more appropriate method. In addition to a greater capital investment for an atomic absorption spectrometer, adequate ventilation is necessary, and also a regular gas cylinder supply.

From an analytical point of view, particulate samples from one work-place will be of relatively constant (and known) composition. Inter-ferences, which limit the use of ultraviolet/visible spectrometry for samples of unknown composition, can be readily assessed.

SAQ 7.3b

> What feature of the sampling and analysis of atmospheric particulates could account for the variety of techniques used for analysis of low concentrations of metals?

Response

The small sample masses of atmospheric particulates (milligrams or below) may mean that you are working close to the limits of detection of the available techniques. The limits of detection of each technique are different for each element (Fig. 7.3a) and so the most appropriate technique may differ for each analysis. In other areas of environmental analysis the sample size may not be such a restriction and pre-concentra-tion may be used to decrease the lower limit of detection.

SAQ 7.4a

> What criteria would you use to choose the analytical technique for several metal ions in a particulate sample?

Response

With such a general question one cannot put the following criteria in any rank order, but they should include the following:

1. Ease of solubility of the analyte. If the analyte is soluble in water or dilute acid, solution analytical techniques are usually the most con-venient to use.

2. Number of elements being analysed. You should re-read the description of the techniques to determine which are most suitable for multiple element analysis.

3. Availability of equipment. Many of the solid-state techniques will only be found in laboratories dedicated to solid-state analysis.

4. Sensitivity. Often you will be working close to the limits of detection of the methods. The most sensitive technique will differ for each element.

5. Compliance with specified method. Some legislation requires the use of a specific procedure for the analysis. Other legislation accepts that alternative techniques may be used if they have suitable accuracy and reliability for the application. The validation of an alternative method may, however, be a long and costly process.

8.   Ultra-trace Analysis

OVERVIEW

This part of the book brings together the ideas you have learnt for the analysis of trace organic pollutants and extends them to analysis at concentrations of ng kg^{-1} or below.

8.1.   INTRODUCTION

So far our discussions have concerned methods using instruments which are commonly available in general-purpose laboratories. Although you may not believe it when you first try to handle concentrations at μg l^{-1} or μg kg^{-1} levels, these methods can be readily performed by skilled analysts. This final section lowers the concentration range studied by a factor of 10^3 or more. We shall start the discussion of the analytical techniques from the point of view of 'how do we modify existing methods to gain the required sensitivity?' We soon find ourselves dealing with instruments which may not be as readily available as those described in the previous sections. The analyst will need to be highly experienced in order to understand the problems when working at such low concentrations. We have now reached the level at which only a few laboratories in any one country have the necessary skill, expertise and facilities to perform the analysis with accuracy and precision.

∏   Why do we need to measure such low concentrations?

1. Concentrations can be greatly increased in an organism compared with the environment in which it is living.

2. Many of the compounds of concern are suspected to have high chronic and/or acute toxicity.

3. Many, but not all of the compounds of concern are thought to be completely man-made and so any detectable concentration gives an indication of environmental contamination.

It is hoped that you did not have any problems with the answer. If you did, you should revise Parts 1 and 2 before proceeding.

What groups of compounds are we discussing?

At the lowest levels of detection, one area of concern is centred around polychlorinated dibenzo-*p*-dioxins (PCDDs). The best known member of the group is 2,3,7,8-tetrachlorodibenzo-*p*-dioxin, which shows very high acute toxicity for some species in laboratory tests. Also of concern is the related polychlorinated dibenzofuran group of compounds (PCDFs), the most toxic member once again being the 2,3,7,8-tetra-chlorinated compound. These compounds are usually found in the environment in complex mixtures containing PCDDs and PCDFs with all possible substitution patterns (Fig. 8.1a).

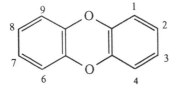 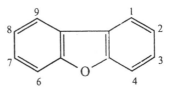

PCDD ring structure   PCDF ring structure

Chlorines may be found in any or all of the substitution positions
1–4 and 6–9

Number of possible PCDDs and PCDFs   = 210
Number of tetrachlorodibenzo-*p*-dioxin isomers = 22
Number of PCDDs and PCDFs which have
chlorines in the 2,3,7,8 substitution positions   = 17

Fig. 8.1a. *Structures and substitution patterns of polychlorinated di-benzo-*p*-dioxins (PCDDs) and polychlorinated dibenzofurans (PCDFs)*

It is not necessary to draw all 210 compounds, but check that there are 17 PCDDs and PCDFs which include the 2,3,7,8-substitution pattern.

The two groups of compounds are formed during the combustion of organic material containing chlorine and are also found as contaminants in some chlorinated chemical products. Combustion sources include chemical and municipal incinerators, coal-fired power stations and domestic coal fires. They also appear capable of being produced naturally by forest and moorland fires.

Polychlorinated biphenyls (an example was shown in Fig. 2.3a) are sometimes included under the category of ultra-trace pollutants. They are found in higher concentrations than PCDDs and PCDFs and can be separated as part of the extraction schemes to be discussed later.

There may be occasions when the compounds already discussed as trace pollutants (Part 4) may need to be monitored at ultra-trace levels. However, the following discussion will be restricted to PCDDs and PCDFs as similar techniques may be applied to other groups of compounds.

∏ The solubility of 2,3,7,8-tetrachlorodibenzo-*p*-dioxin in water is 0.019 μg l^{-1} at 25 °C. It is a solid with vapour pressure at normal temperatures of 6.2×10^{-7} Pa (atmospheric pressure is approximately 10^5 Pa). Using considerations in the earlier sections, what can you deduce of its environmental behaviour? What would be the most suitable samples to take for environmental monitoring?

Using a relative molecular mass of 322, the solubility is equivalent to $5.9 \times 10^{-5} \mu$mol l^{-1}. If you refer to Fig.2.3c, correlating solubility in water with bioconcentration factor, you will find that the factor will be extremely high. It is in fact off the scale in Fig. 2.3c. The compound is also likely to accumulate in sediments. There are few investigations of dioxin concentrations in natural water samples (concentrations would be at or below the lower detection limits), but many in sediments and living organisms. The low vapour pressure indicates that the dioxin in the atmosphere will be predominantly in the solid state, and particulate analysis would be of major importance.

An additional property of PCDDs and PCDFs is their strong binding ability to organic material in soils. Relatively high concentrations may be found in contaminated soils owing to the binding preventing dispersion of the material. The soil also prevents photolytic degradation which may occur when the compounds are exposed to sunlight.

Typical concentrations of PCDDs and PCDFs are as follows:

	Rural soils (ng kg^{-1})	Urban air (pg m^{-3})	Sewage sludge (ng kg^{-1})	Human fatty tissue (ng kg^{-1})
Tetrachloro-dibenzo-*p*-dioxins	3–8	<0.02–6.5	<0.01–0.37	3–10
Tetrachloro-dibenzofurans	5–30	<0.02–18.7	<0.01–0.90	3–9

Compare these concentrations with typical values of other organic pollutants in soil and sludges (Sections 5.4 and 5.5), the atmosphere (Part 7) and living organisms (Section 5.3) and you will find that the PCDD and PCDF levels are lower by a factor of 1000 or more.

Since there are 75 PCDDs and 135 PCDFs, their separation and subsequent determination is a formidable task, even disregarding the low concentrations involved and possible interference by large numbers of other components in the sample. The highest toxicity occurs when there are 4–6 chlorines per molecule, with substitution in the 2,3,7,8-positions. Most investigations restrict the analysis to compounds with four or more chlorines.

SAQ 8.1a In the light of the low concentrations involved, suggest why we have to measure each individual PCDD or PCDF rather than the total PCDD or PCDF.

SAQ 8.1a

8.2. ANALYTICAL METHODS

8.2.1. General Considerations

Let us start with our existing knowledge of the analysis of organic
micropollutants.

∏   What were the main stages of the analytical determination?

1. Extraction of the analyte.

2. Separation from interfering compounds by chromatography.

3. Concentration.

4. Analytical separation and determination by gas chromatography,
   using an electron-capture detector for chlorinated compounds.

You should revise Section 4.2 if you have not remembered these steps.

We must now consider modifications to the method to analyse $ng\,l^{-1}$
rather than $\mu g\,l^{-1}$ concentrations. There would appear to be two routes:

1. Increasing the overall concentration factor in the pretreatment stages.

2. Increasing the detection sensitivity.

There may be room for a small improvement in overall sensitivity by method 1 but this will be insufficient on its own. Note, for instance, the final extract volume (1 ml) given in the DDT analytical method in Section 4.2. There could be at least a tenfold decrease in this volume, with a corresponding increase in sensitivity. The possibility of increasing the sample size is limited by physical considerations of handling large samples under clean conditions. A disadvantage of this method is that any impurities which are carried through the extraction scheme will also tend to concentrate.

Method 2 could lead to a more substantial increase in overall sensitivity, particularly if selective detection of the desired analyte is included

8.2.2. Factors Affecting Detection Sensitivity

The determining factor for this is often the 'random' baseline fluctuations in the chromatogram due to unresolved minor peaks. These peaks could be due to components not removed in the sample pretreatment.

∏   What approaches do you think could be used to minimise this effect?

1. Increasing the chromatographic resolution of the column. This would decrease peak overlap and more readily allow minor peaks to be distinguished from the true baseline. High-resolution capillary columns are essential for ultra-trace analysis.

2. Increasing pretreatment to increase the removal of minor components. This will increase analytical time (which can already be of the order of 24 h). Each additional step increases the possibility of sample loss or contamination.

3. Changing to a more selective detector which would not respond to the minor peaks. Most current investigations use some form of mass spectrometric detection. This may allow a simplified pretreatment but can increase, sometimes almost prohibitively, the cost of instrumentation and hence decrease the number of laboratories which can be equipped to perform the analysis.

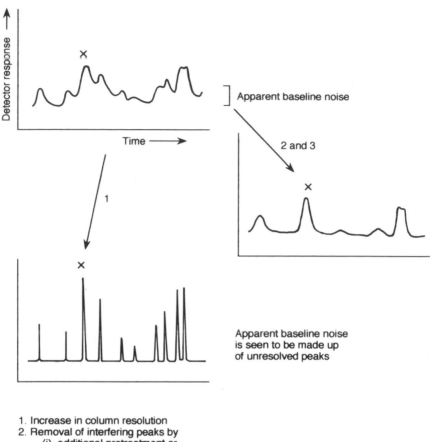

1. Increase in column resolution
2. Removal of interfering peaks by
    (i) additional pretreatment or
    (ii) use of a selective detector
The limit of detection is often defined as
    Peak height = 2.5 × baseline noise

Fig. 8.2a.   *Methods of increasing detection sensitivity*

The effect of the three approaches is shown in Fig. 8.2a.

The wide variety of analytical methods found in the literature result from different emphases being placed on the last two methods. In order to understand this we now need to examine how mass spectrometers may be used as selective GC detectors.

8.2.3. Mass Spectrometric Detection

You should have already come across mass spectrometry as a method of identifying organic compounds.

∏   Briefly describe this technique and how it is used to identify organic compounds.

The compound is ionised under high vacuum, often using electron impact. In the process the molecules fragment. The ions produced are focused into a beam, accelerated and then separated according to their masses (or more precisely their mass/charge ratios, m/z). High-resolution mass spectrometers separate the ions using both magnetic and electrostatic fields (Fig. 8.2b). Low-resolution spectrometers, which are commonly found in bench-top gas chromatograph–mass spectrometer (GC–MS) systems, tend to use electrostatic quadrupole separation.

The peak with the highest mass/charge ratio is usually (but not always) from the unfragmented ion. This can be used to confirm the relative molecular mass of the compound. The fragmentation pattern (Fig. 8.2c) can give an indication of the chemical groups in the molecule. Under favourable circumstances, the molecular structure can be determined from this. However, a simpler method of identification is to compare the spectrum with that from a pure sample, or from a reference library.

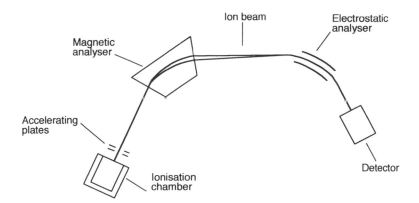

Fig. 8.2b. *Schematic diagram of a high-resolution mass spectrometer*

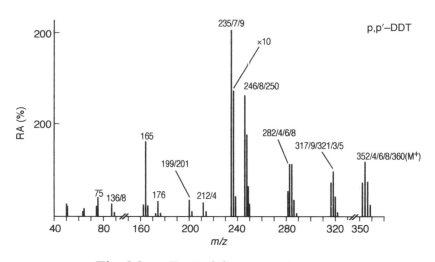

Fig. 8.2c. *Typical fragmentation pattern*

In order to use the mass spectrometer as a universal GC detector, the total ion current is monitored. A typical chromatogram is shown in Fig. 8.2d. When dealing with simple mixtures, the chromatographic peaks may then be identified, and their purity confirmed by the production of a complete mass spectrum for each peak or part of a peak. Even with this simple use of a mass spectrometer you can see how much data can be generated and why the widespread use of GC–MS had to await the development of cheap computer data storage!

One of the problems with the analysis of PCDD/PCDF mixtures is that different species can give identical fragmentation patterns and it is difficult to identify a species on the basis of its mass spectrum alone.

The above application of GC–MS is useful for preliminary chromatographic surveys but still does not use the capability of the spectrometer as a selective detector. The simplest way to do this is to monitor a single ion, this usually being the molecular ion of the compound, i.e. the ion with the same molecular mass as the parent molecule. There is an increase in sensitivity in comparison with total ion current detection since the detector spends all its time monitoring one ion rather than scanning the complete range. The chromatogram produced contains fewer peaks than a total ion current chromatogram, but in a mixture

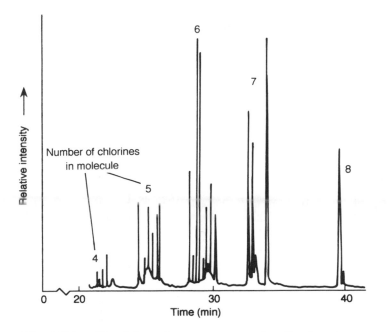

Fig. 8.2d. *Total ion chromatogram of a dioxin mixture*

such as a dioxin extract, the chromatogram may still be complex. If the detector is set at m/z 322, for instance, all 22 tetrachlorodibenzo-p-dioxin isomers will be detected, in addition to ions from other compounds which coincidentally have m/z 322.

Potential interferences in the chromatogram can be detected if fragments are monitored at two or more mass/charge ratios. This is known as selected-ion monitoring. When applied to dioxin analysis the technique makes use of naturally occurring chlorine being found as an approximate 3:1 mixture of ^{35}Cl and ^{37}Cl isotopes. Any molecular fragment containing one chlorine atom will be able to be detected at two mass/charge ratios separated by 2 atomic mass units, corresponding to the ions containing ^{35}Cl and ^{37}Cl. Their intensities should be in the ratio 3:1. If the fragment is not detected at both m/z values, then you have been wrong in your assumption that the fragment contains chlorine. If the relative intensities are not 3:1, and you are certain that there is just one chlorine in the fragment, then this would suggest there is interference from a second ion which coincidentally has an m/z value identical with that of one of the ions.

If the fragment contains more than one chlorine atom, the pattern will become more complex, but still predictable and easily recognisable with experience. The relative intensities of ions containing between one and four chlorine atoms are shown in Fig. 8.2e.

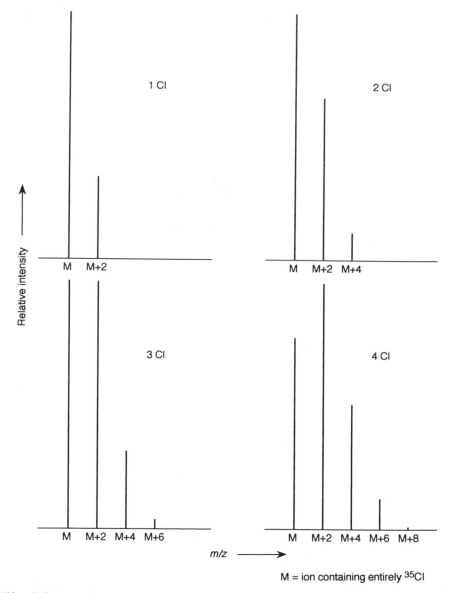

M = ion containing entirely ^{35}Cl

Fig. 8.2e. *Relative intensities of ions containing more than one chlorine atom*

∏   What m/z values could be used to detect the unfragmented 2,3,7,8-tetrachlorodibenzo-p-dioxin ion?

The highest intensity ions will be the following:

$$m/z$$

$$^{12}C_{12}{}^{1}H_4{}^{16}O_2{}^{35}Cl_4 \quad = 320$$
$$^{12}C_{12}{}^{1}H_4{}^{16}O_2{}^{35}Cl_3{}^{37}Cl = 322$$
$$^{12}C_{12}{}^{1}H_4{}^{16}O_2{}^{35}Cl_2{}^{37}Cl_2 = 324$$
$$^{12}C_{12}{}^{1}H_4{}^{16}O_2{}^{35}Cl^{37}Cl_3 = 326$$
$$^{12}C_{12}{}^{1}H_4{}^{16}O_2{}^{37}Cl_4 \quad = 328$$

There will also be a number of lower intensity peaks due to ions containing one or more ^{13}C atoms, rather than all the carbons atoms being ^{12}C. The natural abundance of ^{13}C is 1.08%.

Statistical considerations show that the m/z 320 and 322 ions are the most abundant (relative intensities 77:100) and, in practice, it is just these ions which are normally monitored.

You should note that, although the existence of chlorine isotopes is a considerable advantage for the identification of chlorine-containing fragments, it does increase the possibility of other species being detected at any chosen m/z. Consider p,p'-DDE (Fig. 2.3e). The most abundant molecular ion has m/z 318 and so would not be detected if the spectrometer is set at m/z 320 and 322. There will, however, also be lower intensity peaks at m/z 316, 320, 322 and 324, two of the ions potentially interfering in tetrachlorodibenzo-p-dioxin determinations.

∏   As described so far, it would appear that selected-ion monitoring possesses a major problem for the analysis of dioxin mixtures. What is this?

If the detection is set at, say, m/z 320 and 322, then only the 22 tetrachlorinated isomers will be detected, and not other polychlorinated dioxins. The detector could be set to include additional mass/charge ratios corresponding to the other species, but a compromise would have to be reached as the overall sensitivity decreases with an increase in the number of ions detected.

The problem can be overcome by changing the ions monitored during the course of the elution. This, of course, leads to the requirement for the chromatographic column to separate the mixture into isomer groups. Most columns used for dioxin analysis are able to achieve this group separation but this is usually at the cost of incomplete resolution of some individual isomers. You may wish to look back at Fig. 8.2d to check this statement for a typical chromatographic separation, and to determine possible ion detection sequences which could be used for selected-ion monitoring.

We have now discussed both parts of the most common detection method used in dioxin analysis, namely

selected-ion detection at two or more m/z values;

detection of different isomer groups by change of m/z values monitored throughout the chromatogram.

There are, however, a number of other approaches to selective detection. Two of these will be discussed below.

Other Mass Spectrometric Methods

Although some ions may appear to have identical masses on low-resolution spectrometers, they may often be differentiated using high-resolution instruments. For example, the following ions (molecular ions or fragments) would be detected at m/z 322 on a low-resolution instrument:

	Accurate mass
Tetrachlorinated dioxins	321.8936
DDE	321.9292
DDT	321.9219

These may be selectively detected by high-resolution mass spectrometry owing to slight differences in their accurate masses. Although this is an obvious advantage as the pretreatment may be reduced, most literature methods still use low-resolution mass spectrometry owing to its lower cost and wider availability.

A further development is tandem mass spectrometry (MS/MS) where a single ion (say m/z 322) is subjected to a second fragmentation to confirm the identity of the ion. Initial hopes in the development of the method were that such a method would completely remove the necessity for a pretreatment stage, even for dioxins in complex sample matrices, and hence produce a considerable analytical time saving. Pretreatment has, however, still been found to be necessary for some samples. The sensitivity can also be lower than high resolution MS.

8.2.4. Quantification

This is normally performed by addition of known amounts of internal standards to the sample before extraction. The method will compensate for sample losses in the clean-up stage, assuming that the losses of the standard are identical with those of the analyte, and will also ensure that the determination is independent of any variations in the sensitivity of the spectrometer. Fully substituted ^{13}C isotopically labelled compounds are often used.

Ideally one ^{13}C internal standard should be added for each compound to be determined. This is, however, not generally practicable. It is common practice to use just one standard for each isomer group, and this would normally be the compound containing the 2,3,7,8-substitution pattern. Isotopically labelled derivatives are available for all 17 PCDDs and PCDFs which contain this substitution pattern.

Concentrations of non-2,3,7,8-isomers can be calculated if their response with respect to the 2,3,7,8-isomer can be determined using reference samples of pure material. These may not always be available. In general, however, other compounds are not determined individually, isomer group concentrations (e.g. total tetrachlorinated PCDDs) generally being considered sufficient. Look back at the table in Section 8.1 to find an example of their use. The group concentrations are determined using an average response factor calculated from as many individual isomer response factors as are available.

8.2.5. Quality Control

A second internal standard is often added immediately before injection into the GC–MS system. The standard can be a ^{13}C- or ^{37}Cl-labelled compound. This is for quality control purposes. It allows determination of the recovery of the dioxin over the clean-up stage. A low recovery would give rise to concern over the accuracy of the final results.

The second standard can also be used to provide an estimate of the sensitivity of the detector, which may vary over a period of time. This is an important control feature as most determinations involve operating the instrument close to the limit of detection.

SAQ 8.2a

> The analytical technique for PCDDs and PCDFs from solids includes extraction followed by clean-up and concentration of the extract then GC–MS analysis.
>
> Why is an isotopic internal standard better than, say, a compound which is structurally similar to PCDDs and PCDFs and is not found in the analytical mixture?

SAQ 8.2b Using GC–MS with selected-ion monitoring, how would you set about confirming the identity of a low-intensity chromatographic peak as a particular dioxin?

8.3. A TYPICAL ANALYTICAL SCHEME

The pretreatment is summarised in Fig. 8.3a and the subsequent chromatographic separation in Fig. 8.3b.

The scheme is an example of method where the emphasis is placed on sample clean-up and separation combined with low-resolution mass spectrometry, rather than relying on high-resolution MS techniques.

The scheme is used below as an exercise to test your understanding of the principles of sample pretreatment and the subsequent analytical determination. This will form a suitable conclusion to this open learning material on environmental analysis.

Pretreatment

First, let us compare the scheme with the analysis of DDT as discussed in Section 4.2. You should note the overall similarity of the individual stages.

∏   What are the major differences between them?

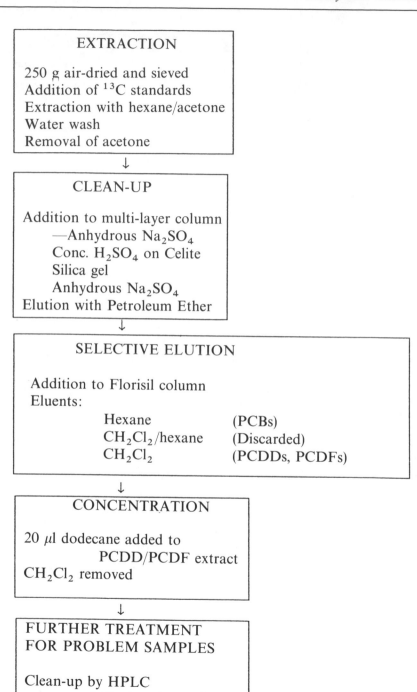

Fig. 8.3a. *A typical pretreatment scheme for soils*

1. The initial chromatographic clean-up for the dioxin analysis uses more than one stationary phase, reflecting the complexity of the extract.

   The number and type of phases used for dioxin extract clean-up vary considerably between literature methods. Acidic and basic silica and alumina columns are in common use. Note here the use of a multi-layer column rather than individual columns, saving analytical time and minimising the possibility of sample contamination or loss.

2. Additional pretreatment for 'problem' dioxin samples.

   HPLC separation using graphitised carbon is used. Graphitised carbon has been found to be highly selective towards planar molecules. It is a straightforward, although time-consuming, operation to determine which sample types need further pretreatment, by comparison of the GC–MS chromatograms for one sample with and without the additional step.

Note, in both steps above, the necessity to minimise the pretreatment time whilst still maintaining the efficiency of the clean-up.

∏ Why do you think there is a change of solvent composition between extraction and clean-up for the dioxin analysis?

The extraction stage uses a hexane/acetone mixture. Acetone is often used as a solvent modifier in extractions from solids to increase the polarity of the solvent and to assist in the penetration of the solvent into the samples (Section 5.3). The first clean-up procedure involves application of the extract on to a chromatographic column and subsequent elution of the non-polar components with a non-polar solvent (light petroleum). The presence of a polar solvent in the extract would lower the efficiency of the chromatographic separation.

In both the DDT and dioxin analytical procedures a second column is required to separate chlorinated species (pesticides, PCBs, etc.).

∏ Why do you think these components were not removed by the first column?

The chlorinated compounds have similar chemical structures. They are all neutral, non-polar, high molecular mass compounds and will have similar chromatographic retention properties. The first column in any clean-up is generally used to remove interferences from compounds with widely different chromatographic properties. The non-polar eluent used will elute the chlorinated species together. Separation of these closely related species will require a second and more selective column with sequential elution of the compounds with a series of solvents of increasing polarity.

The necessity to ensure purity of solvents and cleanliness of apparatus has been discussed in Part 4 and needs to be re-emphasised here. All batches of solvents and reagents need to be checked frequently to confirm a lack of contamination. Pesticide-free or distilled-in-glass grade solvents should be used. Extreme care should be taken with respect to known sources of dioxins. Cigarette smoke and ash can contaminate the laboratory. Extraction thimbles used for solids can be a second source, dioxins potentially being formed by the bleaching process during their manufacture. The thimbles should be pre-extracted with solvent prior to use in the analysis.

Gas Chromatography

The chromatographic column has both to separate the components of the mixture and be compatible with mass spectrometric detection.

∏   From your knowledge of the analytical problem, what can you say about the type of column required.

1. First, in order to resolve the total number of components and to interface with the mass spectrometer, narrow-bore capillary columns are necessary. A programmed temperature gradient will optimise the separations.

2. The stationary phase would have to be compatible with high-temperature operation to elute the lower vapour pressure compounds. A silicone stationary phase would be the obvious choice.

Tetra- and pentachlorodibenzo-*p*-dioxins and furans:
    5 m × 0.2 mm i.d. BP5 (medium polarity) capillary column
    plus
    50 m × 0.2 mm i.d. RSL 950 or CP-SIL 88 (highly polar) capillary
    columns
    Oven temperature gradient 170–240 °C
Hexa-, hepta- and octachlorodibenzo-*p*-dioxins and furans:
    50 m × 0.2 mm i.d. BP5 capillary column
    Oven temperature gradient 170–290 °C
    Elution time 60 min

Fig. 8.3b. *Chromatographic conditions used in the HMSO analytical method. Reference:* Determination of Polychlorinated Biphenyls, Polychlorinated Dibenzo-*p*-dioxins and Polychlorinated Dibenzofurans in UK Soils. *HMSO, 1989*

3. With a mixture of compounds of varying polarity (according to the degree of substitution and substitution pattern), a medium-polarity stationary phase would be a good first try.

4. For mass spectrometric detection, it would be advantageous for the column to group the eluted compounds into isomer groups. This aids peak identification in addition to allowing a fairly simple ion sequence to be used for detection.

The separation of all 210 dioxins and furans and their division into separate isomer groups are exacting demands for a single capillary column. You may not be surprised to find that two columns and multi-dimensional GC/GC were used in the analytical scheme (Fig. 8.3b).

SAQ 8.3a	Imagine that you were about to analyse a large number of samples by a method such as described in Section 8.3. What features would you include in your scheme to ensure analytical quality throughout the programme?

SAQ8.3a

SAQ 8.3b   There are at least two areas of uncertainty in the analytical procedures for dioxins in solid samples as exemplified above. What are these?

Summary

Some species (e.g. dioxins and related compounds) have such a great ability to bioaccumulate and such a high degree of toxicity that monitoring their presence at $ng\,l^{-1}$ or $ng\,kg^{-1}$ concentrations is necessary. The analyses not only require highly sensitive and selective instrumentation but also a large degree of analytical skill and expertise. Gas chromatography–mass spectrometry is most often used. The technique has been discussed along with the necessary clean-up and concentration stages.

Objectives

You should now be able to:

- understand the problems presented by analyses of organic compounds (particularly polychlorinated dibenzo-*p*-dioxins and related compounds) at concentrations of $ng\,kg^{-1}/ng\,l^{-1}$ or below;

- appreciate and be able to describe the use of GC–MS for the analysis of these compounds, including the application of isotopes and isotopic standards to quantification and quality control.

SAQs AND RESPONSES FOR PART EIGHT

SAQ 8.1a | In the light of the low concentrations involved, suggest why we have to measure each individual PCDD or PCDF rather than the total PCDD or PCDF.

Response

1. Each of the components will have different physical and chemical properties, which in turn leads to different bioconcentration ability, rates of degradation and toxicity.

2. As you might expect, the relative quantities of each of the compounds will be different from each production source. Under favourable circumstances, estimation of the relative concentrations can give an indication of their likely origin.

SAQ 8.2a

> The analytical technique for PCDDs and PCDFs from solids includes extraction followed by clean-up and concentration of the extract then GC–MS analysis.
>
> Why is an isotopic internal standard better than, say, a compound which is structurally similar to PCDDs and PCDFs and is not found in the analytical mixture?

Response

The assumption with the use of internal standards is that the standard will behave in the extraction identically with the compound being analysed. An isotopically labelled compound would be closer in behaviour than a chemically distinct compound.

A second benefit is that the labelled compound serves for peak identification, an important consideration when you remember the large number of peaks which may be found even in a selected-ion chromatogram.

SAQ 8.2b

> Using GC–MS with selected-ion monitoring, how would you set about confirming the identity of a low-intensity chromatographic peak as a particular dioxin?

Response

1. The peak should occur at the expected retention time for the chromatographic column. It is easy to forget that the mass spectrometer is simply a highly sophisticated detector for the chromatograph and that retention times are a good primary means of identification.

2. The peak should be monitored at two or more m/z values, corresponding to the same molecular fragment with different distributions of ^{35}Cl and ^{37}Cl in the molecule. The relative intensities should correspond to the expected statistical distribution. A complete mass spectrum could be used to attribute the peak to a dioxin or dibenzofuran rather than an impurity. However, the fragmentation patterns

of the dioxins and furans are often too similar to allow positive identification of an individual member of the groups.

3. The peak should be considered genuine only if it is at least 2.5 times greater than the background noise. Below this intensity there is a possibility that the 'peak' may simply be part of the background.

SAQ 8.3a | Imagine that you were about to analyse a large number of samples by a method such as described in Section 8.3. What features would you include in your scheme to ensure analytical quality throughout the programme?

Response

You should include in your programme:

blank determinations of all batches of reagents used;

analysis of standards from National Laboratories;

replicate analyses, which may be unmarked ('blind') replicates;

repetition of one unknown sample throughout the sequence:

frequent checks on:

purity of reagents throughout the programme;

the recovery standards in the pretreatment;

the resolution of the GC column.

These features are little different from those which you would include in any analytical scheme, but for PCDD and PCDF analysis there are severe limitations on how many standards and replicates may be included owing to the laboratory time required for each sample. Remember that it takes 1 h for each GC analysis, in addition to the time taken per sample in the pretreatment stage.

SAQ 8.3b

> There are at least two areas of uncertainty in the analytical procedures for dioxins in solid samples as exemplified above. What are these?

Response

The first is common to all analyses where there is extraction from a solid. There is always the uncertainty that the extraction is complete. The extraction efficiency of the compound within the sample matrix may also differ from that of the internal standard.

The second arises from the impracticability of using internal standards for all 210 PCDDs and PCDFs, and the uncertainty involved in the determination of average response factors. The practical limit is often seen as one standard per isomer group, usually the compound including the 2,3,7,8-substitution pattern.

Units of Measurement

There is a bewildering array of units of measurement in common use in environmental chemical analysis. Units used to describe the composition of water, atmospheres and solids have developed independently and in each of these areas there may be more than one system in frequent use. SI units (Système Internationale d'Unités)—based on the definition of five basic units: metre (m); kilogram (kg); second (s); ampere (A); mole (mol); candela (cd)—which had been recommended by many international scientific bodies are unfortunately not often used.

The approach used in this volume is to use the units most commonly found in the environmental literature but wherever possible to choose units based on

mass of analyte/unit volume	Water
	Atmosphere
mass of analyte/unit mass	Solids

Typical units would then be

Water	$mg\ l^{-1}$
	$\mu g\ l^{-1}$
Atmosphere	$mg\ m^{-3}$
	$\mu g\ m^{-3}$
Solids	$mg\ kg^{-1}$
	$\mu g\ kg^{-1}$

The alternative system of units sometimes found in the environmental literature based on parts per million (ppm), parts per billion (ppb) and sometimes parts per trillion (ppt), is avoided wherever possible owing to possible ambiguity in the interpretation of the units. As shown below, different definitions are used for the terms in liquids or solids and gas analysis.

For liquids (and solids):

ppm = parts per million (mass/mass)
   $= mg\ kg^{-1}$
   $\approx mg\ l^{-1}$ (assuming density of sample $\approx 1\ g\ ml^{-1}$)

Similarly:

ppb $\approx \mu g\ l^{-1}$
ppt $\approx ng\ l^{-1}$

For gases:

ppm = parts per million (volume/volume)
   $= \mu l\ l^{-1}$
ppb $= nl\ l^{-1}$
ppt $= pl\ l^{-1}$

Note should also be taken that billion and trillion follow US rather than UK usage, i.e.

1 billion $= 10^9$
1 trillion $= 10^{12}$

Table 1 shows some symbols and abbreviations commonly used in analytical chemistry; Table 2 shows some of the alternative methods for expressing the values of the physical quantities and the relationship to the value in SI units; Table 3 gives prefixes for SI units.

More details and definition of other units may be found in D. H. Whiffen, *Manual of Symbols and Terminology for Physicochemical Quantities and Units*, Pergamon Press, 1979.

Table 1 *Symbols and Abbreviations Commonly Used in Analytical Chemistry*

Å	ångström
$A_r(X)$	relative atomic mass of X
A	ampere
E or U	energy
G	Gibbs free energy (function)
H	enthalpy
J	joule
K	kelvin $(273.15 + t\,°C)$
K	equilibrium constant (with subscripts p, c, therm, etc.)
K_a, K_b	acid and base ionisation constants
$M_r(X)$	relative molecular mass of X
N	newton (SI unit of force)
P	total pressure
s	standard deviation
T	temperature/K
V	volume
V	volt $(\text{J A}^{-1}\,\text{s}^{-1})$
$a, a(A)$	activity, activity of A
c	concentration/mol dm^{-3}
e	electron
g	gram
i	current
s	second
t	temperature/°C
bp	boiling point
fp	freezing point
mp	melting point
$\approx$	approximately equal to
$<$	less than
$>$	greater than
e, $\exp(x)$	exponential of x
$\ln x$	natural logarithm of x; $\ln x = 2.303 \log x$
$\log x$	common logarithm of x to base 10

Table 2 *Summary of Alternative Methods of Expressing Physical Quantities*

1. Mass (SI unit: kg)

g $= 10^{-3}$ kg

mg $= 10^{-3}$ g $= 10^{-6}$ kg

μg $= 10^{-6}$ g $= 10^{-9}$ kg

ng $= 10^{-9}$ g $= 10^{-12}$ kg

2. Length (SI unit: m)

cm $= 10^{-2}$ m

mm $= 10^{-3}$ m

μm $= 10^{-6}$ m

3. Volume (SI unit: m³)

l $= dm^3 = 10^{-3}$ m³

ml $= cm^3 = 10^{-6}$ m³

μl $= 10^{-3}$ cm³ $= 10^{-9}$ m³

4. Concentration (SI unit: mol m⁻³)

M $= $ mol $l^{-1} = $ mol dm$^{-3} = 10^3$ mol m^{-3}

Water

mg l^{-1} $= 10^{-3}$ g dm$^{-3} \approx$ mg kg$^{-1} = $ ppm (m/m)

μg l^{-1} $= 10^{-6}$ g dm$^{-3} \approx \mu$g kg$^{-1} = $ ppb (m/m)

ng l$^{-1} = 10^{-9}$ g dm$^{-3} \approx$ ng kg$^{-1} = $ ppt (m/m)

Table 2

Solids

$$mg\ kg^{-1} = 10^{-6}\ g\ g^{-1} = ppm\ (m/m)$$

$$\mu g\ kg^{-1} = 10^{-9}\ g\ g^{-1} = ppb\ (m/m)$$

$$ng\ kg^{-1} = 10^{-12}\ g\ g^{-1} = ppt\ (m/m)$$

Atmosphere

$$mg\ m^{-3} = 10^{-6}\ g\ l^{-1}$$

$$\mu g\ m^{-3} = 10^{-9}\ g\ l^{-1}$$

$$ppm\ (v/v) = 10^{-6}\ m^3\ m^{-3} = cm^3\ m^{-3}$$

$$ppb\ (v/v) = 10^{-9}\ m^3\ m^{-3} = 10^{-3}\ cm^3\ m^{-3}$$

5. Pressure (SI unit: $N\ m^{-2} = kg\ m^{-1}\ s^{-2}$)

Pa	$= N\ m^{-2}$
atm	$= 101\,325\ N\ m^{-2}$
bar	$= 10^5\ N\ m^{-2}$
Torr	$= mmHg = 133.322\ N\ m^{-2}$

6. Energy (SI unit: $J = kg\ m^2\ s^{-2}$)

cal	$= 4.184\ J$
erg	$= 10^{-7}\ J$
eV	$= 1.602 \times 10^{-19}\ J$

Table 3   *Prefixes for SI Units*

Fraction	Prefix	Symbol
10^{-1}	deci	d
10^{-2}	centi	c
10^{-3}	milli	m
10^{-6}	micro	μ
10^{-9}	nano	n
10^{-12}	pico	p
10^{-15}	femto	f
10^{-18}	atto	a

Multiple	Prefix	Symbol
10	deca	da
10^2	hecto	h
10^3	kilo	k
10^6	mega	M
10^9	giga	G
10^{12}	tera	T
10^{15}	peta	P
10^{18}	exa	E

Index

An asterisk next to a page reference indicates mention of an analytical method for the compound or ion.